D0982024

RARE

RARE

The High-Stakes Race to Satisfy Our Need for the Scarcest Metals on Earth

KEITH VERONESE

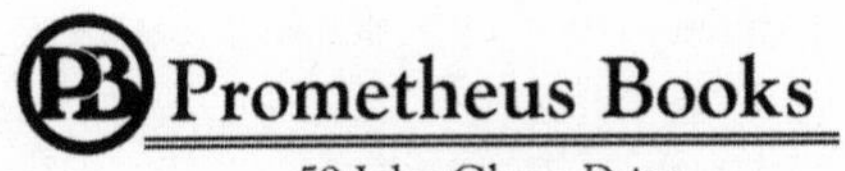

59 John Glenn Drive
Amherst, New York 14228

Published 2015 by Prometheus Books

Cover image of niobium (top) © Visuals Unlimited
Cover image of titanium (bottom) © Getty Images
Jacket design by Nicole Sommer-Lecht

Inquiries should be addressed to
Prometheus Books
59 John Glenn Drive
Amherst, New York 14228
VOICE: 716–691–0133
FAX: 716–691–0137
WWW.PROMETHEUSBOOKS.COM

19 18 17 16 15   5 4 3 2 1

Library of Congress Cataloging-in-Publication Data

Veronese, Keith, author.
Rare : the high-stakes race to satisfy our need for the scarcest metals on Earth / by Keith Veronese.
pages cm
Includes bibliographical references and index.
ISBN 978-1-61614-972-7 (hardback) — ISBN 978-1-61614-973-4 (ebook)
1. Rare earths. I. Title.

QD172.R2V43 2015
338.4'7669291—dc23

2014026931

Printed in the United States of America

For Carla, the best person I know.

CONTENTS

CHAPTER 1

MAN AND METAL

MAN AND METAL

We are surrounded by a cacophony of metals. Aluminum cans store sweet, caffeine-filled beverages that give an afternoon boost. Iron support structures weave throughout every modern building, and our pockets jingle with nickel and copper coins. The rare metals gold and platinum may be a little less common in our day-to-day lives, but they play important roles in electronics, internal combustion engines, and, of course, jewelry. In the past three decades, industrial needs have led to the use of a number of lesser known, rare metals to advance consumer electronics, health care, communications, and the defense industry.

These "new" metals are often fugitives from the bottom rows of the periodic table—metals with names like niobium, tantalum, and rhodium. These elements were likely overlooked in your high school or college chemistry class, but they play a litany of roles in the technological growth of society, as engineers, chemists, and physicists push the boundaries of the exciting and possible.

Humankind's past is littered with methods of using metals to fashion weapons and tools, with the earliest notations appearing six thousand years ago. While metals may vary in scarcity, this brotherhood shares a number of basic characteristics: they are overwhelmingly malleable and act as superb-quality conductors of electrons, for example. Our word *metal* derives from the Greek word *metallon*. The word loosely translates to "quarry" or "mine" and was often used in reference to bustling gold- and silver mining operations in ancient Greece. This is quite the adequate name, one defined by its source, with metals often requiring a phenomenal amount of work to pry from Earth's grip.

Scientifically, metals are known for a common set of properties. Almost all metals have the ability to transmit electricity and heat—very useful properties in the world of electronics. Most metals can be easily bent and molded into intricate shapes. As a nice bonus, most metals are resistant to all but the most extreme chemical reactions in the outside environment, with the added stability increasing their usefulness. A very apparent exception to this stability, however, is the rusting of iron, a natural process that occurs as iron is exposed to oxygen and water over time in junkyards, barns, and elsewhere.

Is a particular metal hard to find because there is a limited amount, is it simply difficult to retrieve, or does technological demand outpace supply? The acquisition difficulty is likely due to a combination of all these reasons with a dash of nostalgic value to top things off for good measure, particularly for rare metals like gold and platinum that have served as status symbols for thousands of years.

ONE IN A BILLION

Determining a definitive amount of an individual element is a tricky endeavor. If you dive into a physics textbook, you are greeted with estimates of percentage makeup of the universe, solar system, and galaxy by element. If humankind has yet to set foot on Mars, how do we have any idea what percentage of elements make up other galaxies? These estimates are made through a combination of techniques and theories taken from physics, chemistry, and astronomy, relying on high-tech instrumentation to obtain data—after all, several-hundred-million-dollar telescopes are not just for pretty pictures. By looking for fluctuations in the way light bounces back based on what is known regarding how individual atoms of a given element reflect light, rudimentary abundance estimates are made.

Measuring the amount of a given metal on Earth is bit easier but is still an astounding feat. The abundance of metals within the planet's crust is often reported in arcane "parts per" notation. These parts-per-million (and often parts-per-billion) values are attained through painstaking analysis of large amounts of rocks taken from the earth's crust. Ideally, they represent an average amount should all the metals be uniformly distributed throughout the planet—which, as we know, they are not.

Parts per million (or billion), when used without any context, is a horribly obscure measurement that plagued my afternoons spent in Analytical Chemistry Lab. A notation traditionally used to communicate the amount of a contaminant within water or air, parts per million takes a very tiny amount and casts it against a reasonable background. Due to its nature, "parts per" is a dimensionless measurement—it lacks units. These measurements describe a quantity, but not in relation to a known commodity as measurements like fifty miles per hour, ten feet, or one hundred kilometers do.

Thanks to its analytical anonymity, a "parts per" measurement can be used to describe just about anything. By the time you finish the end of this sentence, you have used about two parts per billion of your life—about four and a half seconds of the average human life-span. This sort of measurement is a great way to obscure the facts—a local newscast might state there are three parts per billion of a toxic chemical like lead in the water supply. This is also the way the amount of rare earth metals is often reported. For example, platinum, a scarce, precious metal, exists in four parts per billion of Earth's crust—only four out of a billion atoms within the crust are platinum. This is an extremely small amount. To put the amount of platinum on Earth in an easier-to-visualize light, imagine if one took all the platinum mined in the past several decades and melted it down; the amount of molten platinum would barely fill the average home swimming pool.

Silver, a metal many use on a daily basis to eat with, exists at only a twenty-parts-per-billion value—twenty out of every billion atoms on the planet are silver. Remind your significant other of that fact the next time you go jewelry shopping, and you might save some cash. Osmium, rhenium, iridium, ruthenium, and even gold exist in smaller quantities, much less than one part per billion, while some are available in such small concentrations that no valid measurement exists.

On the extreme end of the scarcity spectrum is the metal promethium. The metal is named for the Greek Titan Prometheus, a mythological trickster who is known for stealing fire from the gods. Scientists first isolated promethium in 1963 after decades of speculation about the metal. Promethium is one of the rarest elements on Earth and would be very useful if available in substantial amounts. If enough existed on the planet, promethium could be used to power atomic batteries that would continue to work for decades at a time. Estimates suggest there is just over a pound of promethium (the most recent estimates suggest five hundred and eighty-six grams) within the crust of the entire planet. When the density of the metal is accounted for, this is just enough of the metal to fill the palm of a kindergartner's hand.

We all know gold and platinum have value, and several other rare metals you might not have heard of, including palladium, rhodium, and osmium, are highly sought after and traded as commodities. But what if an element is on the opposite end of oxygen in the supply-and-demand battle? Let's say there are extremely small quantities of the element, but there isn't enough for the element to be experimented with or put into mainstream use. This is the case with some rare metals, like promethium, one so rare that it lacks an established use. As we will soon learn, lacking a use for promethium is not a loss at the moment, as the total amount of the metal estimated to exist on Earth is astonishingly small.

PULLED INTO THE CORE

Why are some metals spread throughout Earth's crust in an unevenly distributed manner? There are several theories, most of them stemming from a sort of chemical attraction between metals. A number of the rare but extremely useful metals we will be talking about are siderophiles. *Siderophile* is an odd word, meaning "iron loving," and like metallon, it is of Greek origin. Osmium, gold, palladium, and platinum are four of the twelve metals classified by scientists as siderophiles: elements that seek out iron and bond with this common metal.

This special attraction to iron explains why so many prized metals are hard to find. Earth's molten core is estimated to be comprised of up to 90 percent iron, leading the elements to sink into the depths of Earth's crust and continually move closer to the planet's iron core over billions of years. At the same time, this drive to the core depletes the amount of the metals available in Earth's crust. The pull poses a problem to mining efforts—a pull to the core prevents the formation of concentrated deposits that would be useful to mine, leading the metals to instead reside in the crust of our planet in spread-out, sparse amounts. Theoretically, there are a lot more of these highly desired siderophiles present; we just don't have the ability to extract and refine them from the extreme depths of the crust and possibly, from the planet's core.

DIGGING DEEP

The mass of Earth is approximately 5.98×10^{24} kilograms. There is absolutely no easy (or useful) way to put a number of this magnitude into a reasonable context. I mean, it's the entire Earth. I could say something silly, like the mass of the planet is equal to sixty-five quadrillion *Nimitz*-class aircraft carriers, each of which weighs ninety-two million kilograms a piece. This comparison might as well be an alien number, as it lends no concept of magnitude.

The overwhelming majority of Earth's crust is made of hydrogen and oxygen. The only metals present in large amounts within the crust are aluminum and iron, with the latter also dominating the planetary core. These four elements make up about 90 percent of the mass of the crust, with silicon, nickel, magnesium, sulfur, and calcium rounding out another 9 percent of the planet's mass. Making up the remaining 1 percent are the one hundred–plus elements in the periodic table, including a number of quite useful, but very rare, metals.

What is easier to understand are reports of the percentages and proportion of metals and other elements that reside on the surface of the planet and just below. At the moment, Earth's crust is the only portion of the planet that can be easily minded by humans. Deposits of rare metals, including gold, are found under the surface of the planet's oceans, but these deposits are rarely mined for a number of reasons. These metals often lie within deposits of sulfides, solid conjugations of metal and the element sulfur that occur at the mouth of hydrothermal vents. While technology exists that allows for the mining of deep-sea sulfide deposits, extremely expensive remotely operated vehicles are often necessary to recover the metals. Additionally, oceanic mining is a politically charged issue, as the ownership of underwater deposits can be easily contested.

As technology advances, underwater mining for rare metals and other elements will become more popular, but, for the moment, due to cost and safety reasons, we are restricted to the ground beneath our feet that covers about one-third of the planet. Earth's crust varies in thickness from twenty-five to fifty kilometers along the continents, and so far, humankind has been unable to penetrate the full extent of the layer. The crust is thickest in the middle of the continent and slowly becomes thinner the closer one comes to the ocean.

So what does it take to dig through the outer crust of our planet? It takes a massive budget, a long timescale, and the backing of a superpower, and even this might not be enough to reach the deepest depth. Over the course of two decades during the Cold War, the Soviet Union meticulously drilled to a depth of twelve kilometers into the crust of northwest Russia's Kola Peninsula. No, this was not part of a supervillain-inspired plan to artificially create volcanoes but was rather an engineering expedition born out of the scientific head-butting that was common during the Cold War. The goal of this bizarre plot? To carve out a part of the already thin crust north of the Arctic Circle to see just how far humans could dig along and to see exactly how the makeup of the outer layer of the planet would change. Work on the Kola Superdeep Borehole began in 1970, with three decades of drilling leaving a twelve-kilometer-deep hole in the Baltic crust, a phenomenal depth, yet it penetrated but a third of the crust's estimated thickness. As they tore through the crust in the name of science and national pride, the team repeatedly encountered problems due to high temperatures. While you may feel cooler than ground-level temperatures in a basement home theater room or during a visit to a local cavern, as we drill deep into the surface, the temperature increases fifteen degrees Fahrenheit for every one and a half kilometers. At the depths reached during the Kola Borehole expeditions, temperatures well over two hundred degrees Fahrenheit are expected. The extremely hot temperatures and increased pressure led to a series of expensive mechanical problems, and the project was abandoned.

The Kola Superdeep Borehole is the inspiration for the late 1980s and 1990s urban legend of a Soviet mission to drill a "Well to Hell," with the California-based Trinity Broadcasting Network reporting the high temperatures encountered during drilling as literal evidence for the existence of hell. The Soviet engineers failed to reach hell, and they also failed to dig deep enough to locate rare earth metal reserves.

At the moment, we simply lack the technology to breach our planet's crust. The Kola Borehole fails to reach the midpoint of the crust, with at least twenty more kilometers of drilling to go at the time the project was shut down in 1992.

Although Earth's crust holds a considerable amount of desirable metals, if the metals are not in accessible, concentrated deposits, it is usually not worth the cost it would take for a corporation to retrieve them (at least until scarcity and demand elevate desire). The composition of metals within the planet's crust is not uniform, unfortunately, further dividing the world's continents into "haves" and "have nots" when it comes to in-demand metals.

NATIVE AND BOUND

Some metals, like aluminum, are extremely applicable to modern life, with large supplies of the metal retrievable and made into a usable form with a reasonable amount of effort. Not all metals are easy to find or can be isolated in a pure form, adding increased difficulty and cost to the process of mining for rare metals.

Although the metal is abundant and has been shaped and fashioned by humans for thousands of years, copper is very hard to isolate from the crust in a pure form. Bronze, a combination of copper with tin, was sufficient for our ancestors to make weapons and tools, but purer forms of copper and other metals are necessary for the varied number of modern uses. Copper is found within the mineral chalcopyrite. To isolate pure copper from chalcopyrite calls for a work-intensive process that involves crushing a large mass of chalcopyrite, smelting the mineral, removing sulfur, a gaseous infusion, and electrolysis before 99 percent pure, usable copper is obtained. Aluminum, a metal so common it is used to make disposable containers for soft drinks, undergoes a similar process before a form that meets standards for industrial use is obtained.

Due to the difficulty in purifying aluminum found in ore, the metal became extremely prized and more valuable than silver in the eighteenth and nineteenth centuries. Thankfully, there are a variety of ways to separate a desired metal from a backdrop of other metals. In 1886, a chemist in his early twenties ended a five-year academic journey when he discovered a simple but unique process to separate aluminum from ore impurities using electricity. Charles Martin Hall, a professor at Oberlin College, would continue a career as an educator and spend several years teaching in Imperial Japan, but along the way he formed an aluminum-processing company to make use of his newfound process. In time, the company became the industrial giant Alcoa, with Hall serving as vice president. Thanks to this process, aluminum metal went from an expensive, coveted metal in its pure elemental form to a readily available and extremely useful metal akin to iron. Thanks to Hall's process, use of the metal increased exponentially. Hall's method also allowed for supply to keep up with demand, and the metal became relatively inexpensive in a matter of a few years.

Due to its existence as a native metallic element, the process used to purify gold is rather simple—it is, ironically, Bronze Age technology. When found, gold is already in a rather pure form, with little processing needed to prepare the metal for use. To purify gold, samples from various sources are melted together to a temperature just below two thousand degrees Fahrenheit. Picking the proper container is important—you don't want to select a metal or other construct that will deform at extreme temperatures; a ceramic or charcoal-based container is best. The insecticide boric acid is added to increase purity by causing contaminates to remain solid as the pieces of gold are melted. In time, the contaminants aggregate as a thin layer on the top, where they can be skimmed off and the remaining purified gold poured into a form for cooling. The purification of gold is a rather easy process, quite different from the work-intensive steps required to purify the much more prevalent copper. Purifying gold requires only a sample, a proper container, a heat source, and the ability to keep your cool while handling molten metal. The high purity of gold in its initial form no doubt led to a portion of humanity's desire for the metal, allowing it to be readily identified when found.

While many of these metals are scarce, there is one major bright side. Most of the rare earth metals are nonvolatile; they are elements content to be as they are. Thanks to this nature, these elements can be recovered and repurposed—just as the jewelry sold to a gold broker is melted down and resold—so that the usefulness of a finite quantity of these metals can be extended.

Adventurous modern gold "prospectors" purify small pieces of scrap gold scavenged from circuit boards using microwaves. I'm relatively sure I would destroy my microwave or earn myself a trip to the emergency room using this modern smelting twist.

ROCKS INTO SMARTPHONES

The use of exotic metals has become commonplace to improve the activity of existing consumer goods. The piece of aluminum used as part of a capacitor within a smartphone is exchanged for a sliver of tantalum in order to keep up with processor demands, creating an enormous market for the rare metal. Rhodium, ruthenium, palladium, tellurium, rhenium, osmium, and iridium join the extremely well-known platinum and gold as some of the rarest metals on the planet that find regular uses in the medical industry. These rare metals play interesting roles in protecting the environment. A great example is the use of platinum, palladium, and rhodium in catalytic converters, a key component in every automobile built and sold in the United States since the 1970s. Each converter contains a little over five grams of platinum, palladium, or rhodium, but this meager amount acts as a catalyst that turns carbon monoxide into a water vapor and harmless emissions for hundreds of thousands of miles, with the metal unchanged throughout the process.

An extremely recent and highly relevant example of a little-known metal that jumped to the forefront of demand is tantalum. Tantalum is in almost every smartphone, with a sliver in each of the nearly one billion smartphones sold worldwide each year. Prior to this valuable alternative use, tantalum's only major use was as a conducting filament in turn-of-the-twentieth-century lightbulbs, with the metal quickly replaced by tungsten.

Smartphones are just the tip of the proverbial iceberg of innovative uses for these rare metals. Europium is used to create the color red in liquid-crystal televisions and monitors, with no other chemical able to reproduce the color reliably. As copper communication wires are replaced with fiber-optic cable, erbium is used to coat fiber-optic cable to increase the efficiency and speed of information transfer, and the permanently magnetic properties of neodymium lead to its extensive use in headphones, speakers, microphones, hard drives, and electric car batteries.

WILL SUPPLY + DEMAND = CONFLICT?

Metals are governed by the law of supply and demand, like most everything else in our lives. The quixotic pull of supply and demand, however, often makes for some unusual situations. Oxygen is necessary for life—ideally the highest demand. We need it to breathe, and it makes up a considerable portion of life-giving water. However, the abundance of air and what appears to be an unlimited supply renders life-giving oxygen free, except in concentrated forms for medical or industrial use. While a number of rare metals to be covered are not essential to life, they are quite rare, leading to the financial benefit of corporations and manufacture of cutting-edge gadgets, as well as the upheaval of lives and the destruction of villages in some cases.

These relatively unknown but now essential metals are gathered from concentrated deposits in remote areas. Since the use of metal combines the principles of finite resources with the necessity of costly expeditions and retrieval processes, an immediate upsurge in demand often leads to the upheaval of lives across the world. In a world where many countries are quickly becoming "first world" nations, if not politically then technologically, there will be increased competition to acquire tantalum and other high-demand rare metals.

Stories of upheaval and disruption are just one part of the mixed blessing that comes with the discovery of these resources in a region. Conflict metals share a number of parallels with a much sought-after and contested resource: oil. These metals may serve to be the catalyst for a number of political and even military conflicts in the coming centuries. All our heavy metal elements, to which many of the rare metals belong, were born out of supernovas occurring over the past several billion years. These metals, if not recycled or repurposed, are finite resources.

Inside the stories of these rare metals are human trials and political conflicts. In the past decade, the Congo has been ravaged by tribal wars to obtain tantalum, tungsten, and tin, with over five million people dying at the crossroads of supply and demand. Afghanistan and regions near the Chinese border are wellsprings for technologically viable rare metals due to the disproportionate spread of these high-demand metals in the planet's crust. In an interesting move, the United States tasked geologists with estimating available resources of rare metals during recent military actions in Afghanistan.

California, specifically the Mountain Pass Mine within San

Bernardino County, was a leading supplier of rare earth metals in North America well into the 1990s. Mountain Pass, however, was shut down in the early 1990s after a variety of environmental concerns outweighed the additional cost of acquiring the rare earth metals mined there compared to overseas sources.

Since the metals rarely form concentrated deposits, the places in the world that play home to highly concentrated deposits of in-demand metals become the target of corporations and governments. Areas of the world that do happen to contain large deposits of any one of these rare elements—like gold, for example—often experience a quick and violent extraction of these metals. North America witnessed this firsthand during the California Gold Rush of the nineteenth century, as an influx of prospectors attempted to capitalize on imbalance of supply and demand. Discovery of extensive deposits of a not-so-rare metal, tin, was enough to send a part of Africa into a bloody war at the turn of the millennium.

What are these rare metals the world has come to rely on and go to extremes and often deadly lengths to acquire? Let's find out.

CHAPTER 2
WHAT IS RARE?

FLUCTUATING VALUE

During his rule of fourteenth-century BCE Egypt, King Tushratta declared gold to be "more plentiful than dirt" due to its abundance in Northern Africa.[1] Twenty-six hundred years later, Musa I of Mali, ruler of a gold-rich area in modern-day Ghana, made a Muslim pilgrimage with a twist. Musa brought an envoy of twelve thousand slaves to Mecca, each carrying a gold bar, and a select five hundred carrying a six-pound gold staff.[2] During the trek Musa spent and gave away mind-bending amounts of gold, with the numbers of slaves and gold ascribed to the story becoming greater as the tale was repeated over the centuries. While our details of the quantities border on the apocryphal, Musa's pilgrimage had a very real effect on the region, and the influx of gold inflated the value of the precious metal in Cairo, Medina, and Mecca for over a decade. Musa I of Mali visited Mecca on a journey seeking religious inspiration and spreading good will, but he inadvertently caused widespread economic collapse as the value of gold plummeted.

GOLD ABUNDANT AND WORTHLESS?

Circumstances certainly do change. Gold towers above all other metals as the standard for wealth, a position it has held for thousands of years. Silver substantially trails behind gold in worth, with the gap between the two increasing with every passing moment. An ounce of gold in the days of ancient Rome was worth twelve ounces of silver, with this divide becoming a chasm in the intervening fifteen hundred years as an ounce of gold is now worth roughly sixty ounces of silver today. The gap continues to widen due to a combination of tradition, public perception, and scarcity. Our societies are comfortable with gold, silver, and even platinum as standards of worth, but what about other metals?

Up-to-date prices for rare earth metals are not nearly as easy to come across as those for gold, silver, or platinum, and even when found, there is a large learning curve to overcome before one can successfully make a trade. It is not very easy for a private citizen to invest in any one of the seventeen rare earth metals as millions do every year with gold. Due to their relative lower price these rare earth metals and other prized metals are sold by the pound and kilogram. This presents a cumbersome space problem—while a few thousand dollars of gold coins could easily be hidden away in your sock drawer or safety deposit box, the same dollar amount of tantalum will fill a small closet. This barrier to acquisition puts the rare earth metals outside the reach of the run-of-the-mill investor, but these issues are not deterrents to manufacturers, electronics corporations, and countries.

WHAT ARE THE RARE EARTH METALS?

So what exactly is a rare earth metal? Rare earth metal (often abbreviated as REE) is a descriptor given to a group of seventeen different elements that sit side by side, for the most part, on the periodic table. Scandium and yttrium are the only two allowed residence in the main body of the periodic table, with the other fifteen metals comprising the REEs—lanthanum, cerium, praseodymium, neodymium, promethium, samarium, europium, gadolinium, terbium, dysprosium, holmium, erbium, thulium, ytterbium, and lutetium—relegated to a row below the middle of the periodic table, slotted just under the transition metals.

The religious would look at the rare earth metals and deem them holy, elements set apart from the rest of the hoi polloi for an unknown but no doubt lofty reason. In reality, the fifteen rare earths along the bottom are not sacred but are "detained." If the seventeen elements are left in the location deemed proper by their atomic numbers—seated between lanthanum and hafnium on the periodic table—the table becomes unwieldy, immediately doubling in width. Visually, the result would look far less organized than the rectangular grid we are accustomed to, and instead would resemble something like a barbell with oxygen, carbon, and helium on one end and hydrogen, lithium, and sodium on the other. This frankly would not fly with the handful of scientists that guided its creation over the decades, so the fifteen were excised and dropped below.

The fifteen rare earth elements separated and placed below the periodic table are known in historic chemistry circles as the lanthanides. This awkward name is taken from their first member, lanthanum, with each of the remaining fourteen rare earth metals in the row having one more proton than the previous, exhibiting basic properties similar to lanthanum. Overlapping properties in consecutive elements as you look from left to right on the periodic table is an anomaly—elements typically exhibit radical changes with the addition of a proton. Carbon, nitrogen, and oxygen follow each other in the periodic table just as these fifteen do, but the one proton difference between the trio results in wildly different properties. All three are building blocks of life, but each one plays a different role—carbon is used for structural bonds, oxygen is a vital component of air and water, while nitrogen is necessary to every protein and each molecule of DNA—with a single proton difference causing the atoms to bond and behave differently.

The rare earth elements suffer from an unfortunate happenstance—the adjective that takes the spotlight in their name. "Rare" suggests that the metals are nigh impossible to find in any form, but this is not really the case. The more common members of the seventeen metal group are present in Earth's crust in amounts similar to metals we may consider common. For example, the amount of europium, neodymium, ytterbium, holmium, and lanthanum is roughly the same as the amount of copper, zinc, nickel, or cobalt.[3] Simply put, the majority of the seventeen are not rare; they are spread throughout the planet in reasonable amounts. The metals are in high demand and inordinately difficult to extract and process, and it is from a combination of these factors that the seventeen derive their rarity.

RARE VERSUS DIFFICULT TO ACQUIRE

While the seventeen metals may be distributed throughout the planet, finding an extractable quantity is a challenge. The elements are spread so well that they appear in very small, trace quantities—a gram here, a milligram there—in deposits and are rarely, if ever, found in a pure form. Extracting and accumulating useful, high-purity quantities of these seventeen metals is what lends them the "rare earth" name, as their scattered nature spreads them throughout the planet, but in tiny, tiny amounts.

To obtain enough of any one of these seventeen to secure a pure sample, enormous quantities of ore must be sifted through and chemically separated through a series of complex, expensive, and waste-creating processes. The basics of chemical reactions act as a spanner in the works through processing, as the desired metal is lost through side-reactions along the way. Small losses in multiple steps add up quickly, further decreasing the amount of metal available for use.

Why expend so much effort to discover and refine these seventeen rare metals? Many of them are necessary to fabricate modern electronics, metals woven into our everyday lives and used by brilliant scientists and engineers to fix problems and make electronics more efficient at the microscopic level. Think of the seventeen rare earth metals like vitamins—you may not need a large amount of any one of them to survive, but you do need to meet a regular quota of each one. If not, your near future might resemble that of a passenger traveling in steerage from Europe to the New World as you develop scurvy from lack of vitamin C. Yes, we can make substitutes of one of the rare metals for a similarly behaving one on a case-by-case basis, but we need every metal from lanthanum to lutetium, and in sufficient amounts, if we want the remainder of the twenty-first and the upcoming twenty-second centuries to enjoy the progress we benefited from in the twentieth.

Along with the seventeen "not so" rare earths, we will address a handful of other metals with significant global impact. These "scarce" metals, for lack of a better term, include gold, silver, platinum, and other precious metals and are joined by a handful of elements you know of but might not perceive as metals, like uranium and plutonium. Due to radioactivity, it is best that only professionals come in contact with the latter two, and not without a fair amount of lead shielding and a protective gown.

These nondescript rare metals will not be taking home any awards in beauty, with their rather ordinary-looking gray and silver sheen. Their sensitive complexion allows these metals to react freely with oxygen in the atmosphere, leading many of the seventeen to form a "crust" on their surfaces when exposed to the open air. One can remove this layer of tarnish through the use of cleaners and polishing, but not without removing a tiny layer of the metal. The metal is left with an undesirable crust due to an oxidation reaction—the same reaction that turns a 1967 Shelby GT500KR into a rust-covered, pitted shell when the car is forgotten and left to waste away in a barn for decades. Once removed, the patina will return unless the metal is kept sealed away from oxygen, hydrogen sulfide, and a barrage of other small molecules in the environment that can settle on the surface.

ELECTRIC EXTERIORS AND THE CHARACTERISTICS OF RARE EARTH METALS

Despite their rarity and vast array of uses, many metals have a relatively bland, shiny silver exterior. Why are these prized elements given such shabby clothing? The answer to this question is a little heartier than a layer of tarnish easily removed with a two-dollar cream from a jewelry store. Most of the silver and gray metals with which we come in contact put their electrons into organized sets, filling one of the outermost electron orbitals. The electrons in this set, called the "d" orbital by chemists and physicists, are constantly jumping to another set of elections in order to make the entire atom more stable. (Akin to a major league pitcher forgoing a possible free agent payday to sign a six-year contract with his hometown team, atoms are slaves to stability, with this yearning governing almost every chemical bond they form.) In order for our eyes to see any object, light has to bounce off the object and reflect in the direction of our eyes, allowing the light to pass through the lens of the eye and be interpreted by rods and cones in the retina. As light bounces off a sample of a metal rife with electrons jumping from set to set, the electrons in the outer sets of the atoms absorb a small amount of the energy from the incoming light. The remaining light energy is reflected back toward the eye in equal amounts across wavelengths in the visible range of light, with this now spread-out reflection of light becoming the characteristic silver-gray color we ascribe to metals from which our brain and eyes manipulate the information.

So, how does this theory explain one of the more glaring exceptions among the gray-sky backdrop of metals, namely gold? The distance a given electron is jumping in an atom of gold is slightly smaller than it is in the silver-colored metals. The electrons making the shorter jump absorb incoming light a little differently, emitting light with a wavelength corresponding to what we see as yellow.

But what is it about these seventeen metals that make them useful? Reasons vary, but the fifteen elements between lanthanum and lutetium huddled for shelter under the periodic table have a subatomic level of similarity—the fifteen can hide electrons better than the rest of the elements on the periodic table. As mentioned before, when one traces the row of fifteen rare elements from left to right, each additional element along the line gains two subatomic particles—a proton and an electron. Each proton resides in the center of the atom, the nucleus, where they provide a positive attractive force on the clouds of electrons whizzing through the surrounding area. Electrons counteract the positive charge of the protons in the nucleus, with this tug-of-war keeping each individual atom together.

In most cases, extra electrons are deposited on the outside of an atom in distinct sets, where they orbit the nucleus (and the protons within) in arcs that mimic an ellipse or sphere, whizzing past each other like mosquitoes at two thousand kilometers a second. These electrons are moving at nearly six thousand times the speed of sound, yet this is still a tiny fraction of the speed of light—a difficult velocity to comprehend because we simply lack the frame of reference to compare. The electrons circle the nucleus so rapidly that we can view them as effectively surrounding the nucleus in a cloud of electrical charge.

As we look along the rare earth elements, the electrons are added in an organized manner, but how they are added changes depending on the number of electrons already present in each rare earth metal. While it is easier for us to pretend the electrons are spinning around the nucleus in a spherical cloud, the electrons are actually added to different sets (like the "d" orbital we talked about earlier). Each one of these sets represents a different path taken by a group of electrons around a nucleus, akin to the different routes two flocks of birds will take as they depart from New York City and Los Angeles, respectively, as the flocks travel south for the winter. All electrons move around the nucleus, with some moving along a spherical path, others in a football-shaped path, and others in far more complex, lobed routes that annoy the most diligent of students.

The electrons added to the rare earth elements as we move left to right along the row enter a quirky scenario, one that leads to an interesting phenomenon in the rare earth elements. When the new electron is added to its set (one electron for each element after lanthanide), another set of electrons is left unprotected to the positive pull of protons in the nucleus. This is like a football formation wherein a blocking tight end is pulled off the field in favor of an extra wide receiver. If a running play is called, the running back is left without the normal protection from his tight end, increasing the exposure of the runner to the opposing defense. When an electron, the subatomic particle wide receiver in this analogy, is added, another electron in a different area becomes exposed, allowing the protons, our defense, to exert more pull on the exposed electron.

The extra "tug" from protons in the nucleus does not play a role as long as the atom is neutral, but should an electron become dislodged (as often occurs with metals) and an ion is formed, the ion will be smaller in size than normal due the extra pull. When metals form bonds with other atoms and elements, they often do so as ions, with this break from the norm giving the rare earth metals some of their interesting properties.

Because of this phenomenon, ions of the rare earth metals from lanthanum to lutetium grow smaller in diameter from left to right across the row. This is the reverse of typical trends seen in the periodic table, as ions of elements typically become larger across the row. As seen in the rare earth metals, this alteration leads to making ions of these rare earth metals smaller; the electrons traveling along their unique path bestow on the elements interesting magnetic abilities, properties that make rare earth metals particularly sought after for use in electronics and a variety of military applications.

BREAKING DOWN THE MINERALS

Most of our metal supply is obtained through the mining and processing of minerals. "Mineral" is a rather vague term in modern usage. Your breakfast cereal claims to be filled to the brim with vitamins and minerals, but, for our sake, a mineral is a stable, solid material that is found within Earth's crust. Minerals contain a variety of elements, with multiple metals often found in a single mineral deposit. Rocks with a consistently high concentration of a given metal, like magnetite, which has a large amount of iron, are often commonly traded. In a perfect world, all metals would be evenly distributed across the planet and easy to process into a useful form.

Silver and gold, however, tease us. These sought-after precious metals are two examples found in a native metallic form that need very little processing before they can be of use. Native metallic elements are also easy to identify—gold in its native metallic form looks just like gold, with a flicker of yellow sending miners west to California during the nineteenth century. This ease of access and identification has led to the use of silver and gold over several millennia, while metals like tantalum and tungsten are new arrivals on the scene, as technological advancements now allow for their separation from minerals and their subsequent use in a variety of industrial applications.

Mineral deposits differ in the amount of usable metal they contain, with the concentration of metal, ease of extraction, and rarity playing a role in determining how mining operations proceed. Metals are found in a variety of purities, interwoven in a matrix of organic materials and often with other similar metals. Aluminum is found within bauxite deposits, tantalum and niobium are found with the coveted ore coltan, while cerium, lanthanum, praseodymium, and neodymium are found in the crystalline mineral monazite. Recovering a sample from the ground through hours of digging and manual labor is just the first step—before any of these metals can be used, an extensive process of purification is often necessary. This purification process is essential because high levels of purity are necessary for their efficient use.

Five species of minerals dominate our concern in the hunt for rare earth metals: columbite, tantalite, monazite, xenotime, and bastnäsite. We can further reduce this to four species, since columbite and tantalite are often found together in the ore coltan.

Coltan ore contains large deposits of tantalum and niobium, two of the most sought-after rare metals. Central Africa is home to large deposits of coltan, but the fractured nature of the nations in the region and opposing factions have taken the lives of thousands and disrupted countless more as rival groups swoop in to make money off of legal and illegal mining operations in the region.

Raw monazite, xenotime, and bastnäsite are relatively inexpensive. You can buy a rock of the red-and-caramel-colored minerals on any one of a number of websites, with a fingertip-sized piece of monazite or bastnäsite available for the price of a steak dinner at a truck stop diner. Unlike the concentrated deposits of tantalum and niobium in coltan, samples of monazite, xenotime, and bastnäsite minerals hold small amounts of multiple rare earth metals within them. Any rare earths found in the industrial sector or used for commercial products were merely salvaged byproducts of mining easier to acquire and highly desired elements, like uranium.

Sizable deposits of monazite, xenotime, and bastnäsite are found in North America, with a long tradition of exploitation in the United States. Monazite mining projects were scattered throughout North Carolina during the first two decades of the twentieth century, but unfortunately for locals involved in the operations, the boom died quickly. Monazite mining in the state did not fail due to an engineering or safety problem but to a changing economic climate. Burgeoning mining industries in Brazil and India offered monazite at a much lower cost, leaving the North Carolina mines abandoned within a handful of years.

Searching for rare earth metals in monazite brings with it a major problem with the ore—most samples are radioactive. The naturally radioactive metal thorium is a large component of monazite, with the fear of environmental damage, additional economic cost, and employee health concerns acting as barriers to monazite mining operations.

Once a sufficient quantity of any one of these minerals is obtained, there is a long road to tread before the desired metals are pulled from the rocks. Eighteen steps are necessary before monazite can begin to be purified into individual rare earth metals, while bastnäsite requires twenty-four.[4] Some of these steps are simple—crushing and subsequent heating of the raw mineral ore—while others are large-scale chemical reactions requiring highly trained professionals.

The minerals hold tiny amounts of several different rare metals within them. Until recently, carrying out mining operations solely to garner rare earth metals was considered much too expensive. But if the rare earth metals were a useful by-product of other mining and processing efforts, then so much the better.

A great example of this phenomenon is carbonatite, a rock of interest but one less prized than coltan, bastnäsite, xenotime, or monazite. Carbonatite, not to be confused with the fictional chemical carbonite that encased Han Solo in a slab at the climax of *The Empire Strikes Back*, is another mineral of interest to rare earth metal aficionados. Carbonatite is sought for the rich copper content within, with the added bonus of small amounts of rare earth metals that can be teased out as the mineral is broken down.

NOT ALL RARE EARTHS ARE CREATED EQUAL

Nomenclature abounds in our society, as scientists are trained in the practice of breaking down and further organizing their observations whenever possible. The seventeen rare earth elements are often separated further from the periodic table, with the seventeen split into two groups: the light rare earths and the heavy rare earths. As can probably be guessed from the name, mass of the element plays a role in this separation. The light rare earth elements (LREEs) are lanthanum, cerium, praseodymium, neodymium, and samarium, while europium, gadolinium, terbium, dysprosium, holmium, erbium, thulium, ytterbium, lutetium, and yttrium make up the heavy rare earth elements (HREEs). As a general rule, an HREE is harder to find in substantial usable quantities than an LREE, making the heavy rare earth elements more valuable.

The vast majority of the universe is constructed from four elements: an overwhelming amount of hydrogen and helium with a smidgen of oxygen and carbon mixed in for good measure. No atoms of the heavier elements—and definitely not any atoms of our beloved rare earth metals—existed in the reaches of our universe in the moments after the big bang. Atoms of only four elements—hydrogen, helium, lithium, and beryllium—existed in the early seconds of our universe. After hydrogen, helium, oxygen, and carbon, the remaining elements on the periodic table (one hundred and fourteen in all, with the number growing every couple of years as new additions are synthesized) are present in trace amounts in our universe compared to these four. In Earth's crust this distribution changes significantly. Oxygen and hydrogen are still abundant, with silicon and aluminum displacing helium and carbon in the top four. Silicon binds oxygen extremely well, creating the structural foundation for the rocks and minerals wherein many rare metals hide.

Overall, elements that have lower atomic masses (in day-today language, these elements weigh less per atom) are more abundant than atoms with higher atomic masses. Hydrogen atoms (a proton and an electron, so its atomic mass is just over one) and helium atoms (two protons, two electrons, and two neutrons for an atomic mass of four) are two of most abundant in the universe, while the number of elements at the other end of the periodic table with larger masses like gold (seventy-nine protons, seventy-nine electrons, and an average of one hundred and eighteen neutrons for an atomic mass of just under one hundred and ninety-seven) are far less abundant. This trailing phenomenon across the periodic table is part of the answer as to why there are fewer of the heavy rare earths on and within the planet (as well as the rest of the universe) than there are light rare earth elements.

At the moment, 90 percent of the world's current supply of rare industrial metals originates from two countries. The export of raw supplies from these countries is increasingly coming under fire, with the countries championing a movement to convince corporations to move away from the quick monetary gain that exporting raw materials offers and moving toward making a profit by exporting finished consumer electronics. At present, we are seeing the beginning of territorial wars over a far more common resource, fresh water, in the United States and elsewhere in the world. If governments are experiencing difficulties sharing and parceling out water, as we see in ongoing disputes between Alabama, Georgia, and Florida over the Apalachicola-Chattahoochee-Flint River and Alabama-Coosa-Tallapoosa River basins, the quarrels possible over rights to desperately needed metals between non-civil or even warring nations could be frightening.

One country exhibits an overwhelming abundance of rare earth metals, drawing the rest of the world to its door. What country is behind this imbalance, and how did it develop a stranglehold on the world market?

CHAPTER 3

PLAYING THE LONG GAME

Prior to the rise of rare earth mining in China, the United States dominated the world scene. In the decades book-ending World War II, the United States began operations across the country, with the next-largest producer, Australia, producing only a fraction of the US output. The United States remained dominant, maintaining the largest reserve of rare earth minerals in the world for half a century.

What prompted the United States to get serious about rare earths? Before World War II, we relied on a consistent funnel of obscure metals from Brazil and India to meet our supply needs. An effort by both countries to raise prices sent US manufacturers looking inward—a smart decision, since domestic mining efforts were spurred by the US government's search at home and abroad to find uranium for use in nuclear weapons, a search that also turned up a number of rich rare earth deposits. United States–based mining surged again in the mid-1990s to fill the infra-structure demands of the Internet Age, but almost all operations ceased by the end of the decade.

A sudden change in policy took hold in the waning years of the twentieth century, much the same as the low costs that sent the United States to Brazil and India in the first decades of the century. In the 1990s, a number of successful Chinese mining operations began, with their rich supply of high-quality rare earths flooding the global market and driving prices down to near-record lows. The same phenomenon that lured the United States into the rare earths business—rapid price fluctuation—ushered their exit as well. Relying on export streams from China became easier and economically feasible. At the same time, China's power play invited global powers to rely on a single source for these essential materials. The decision to sit on the sidelines may work in a near-sighted scenario, but as it would be in any supply-and-demand situation where a single entity has a near monopoly, such a move will no doubt become less than desirable as time passes.

It is easy to suggest, from a Western perspective, that the United States should play a waiting game and simply allow for reserves in China to be exhausted. Such a mindset is troubling, since every year that passes places an additional delay on the time it would take to ready an existing stateside mine for excavation. Initiating the start-up process at an existing but inoperable mine is not merely a matter of flying equipment and technicians in before management flips a switch—it is an excruciating process compounded in difficulty through the necessity of meeting new environmental legislation and clearing environmental hazards incurred in the absence of operation, measures frustrating an increasingly weary population of taxpayers.

China is not mining these metals merely for a source of income and political clout but first and foremost for its own benefit. China's population is consuming rare earth metals at an astonishing rate. By the year 2016, the population of China is projected to consume one hundred and thirty thousand tons of rare earth metals a year, a number equivalent to the entire planet's consumption in the beginning of this decade.[1] The reasons for this increase are manyfold, but they are in part due to the growing accessibility of high-end technology at a low cost, the rising quality of life for the average Chinese citizen, and a tech-savvy youth culture that rivals any on the planet. We may see a day quite soon when the whole of China's rare earth production is used within its borders, leaving the rest of the world scrambling to meet the needs of their citizens.

After feeling the pull from export limits on rare earths brought on by the Chinese government in 2012, the United States, the

European Union, and Japan pleaded with the World Trade Organization in hopes of forcing China to "play fair." By the time of this plea, the Chinese government had already succeeded in its efforts by giving a handful of countries an idea of what it would be like to face a dramatic decrease in supply. China holds one-third of the planet's rare earth supply, but a vast number of mining and refining operations ongoing within its borders allow China to account for roughly 97 percent of the available rare earth metals market at any given time. Yes, other countries have rare earth metal resources, but they lack the infrastructure or means to put them to use.

The addition of politics into the equation places China in an enviable position of power should a nation or group of nations interfere with the country's interests on any level. Unhappy with the Japanese presence in the South China Sea? Prohibit exports to Japan. Unhappy that a conservative group within the US House of Representatives is calling for a sharp decrease in the number of Chinese nationals allowed to pursue graduate studies in the United States? Play the export card again, but this time on the United States and maybe her allies in Europe and North America, too. Unhappy with just about any political, economic, or social scenario invoking a response greater than public disdain and less than a show of force? Threaten exports again, and keep doing so until you have the ear of the scattered countries of the world, all because their collective citizens want the goods created from China's materials. This comes with an added bonus, since a large percentage of their tax monies spent on military weaponry relies on the same goods that require these rare-metal components, further indebting a sovereign nation.

If the threat of export cessation does not appear significant enough, remember that the United States, the United Kingdom, and the United Nations made use of trade embargoes to guide the military and political behavior of Pakistan, India, and China itself in the early years of the twenty-first century.

The government of China made a decision to play the "long" game in the 1970s—sell large quantities of rare earth metals at very low prices, prices low enough that the rest of the world would flock to this new and inexpensive supplier. Years of allowing outsiders to purchase at below-market value devastated the rare earth mining industry in other parts of the world, and by the turn of the millennium China's mines were the only ones left standing. China set itself, with intention and cunning, at the head of the class of rare earth metals, and the country will be there to stay for the next several decades to satisfy the lack of exploitable supply around the world, as well as the high start-up costs associated with developing a new mine where a worthwhile cache of rare metals exists or reopening an operation shut down in the 1980s or 1990s that closed its doors due to the artificial financial atmosphere created by China's bold move.

While China became the cornerstone of the rare earth metals industry, the United States faltered. Granted, the United States did not necessarily lose its rare earth resources—they remain just under the surface of the ground or in small stores of raw materials waiting until a time of critical use for processing—our efforts to secure and advance this supply went into a deep slumber. China began its ascent to the top of world rare-metal production in 1987, and it has surpassed the United States in production every year since 1992.[2] Corporations that traditionally purchased metal assets owned by North American countries soon traveled across the Pacific to obtain these same metals at a considerably lower cost, mirroring the draw of new discount chain stores on the inhabitants of surrounding neighborhoods. Cost and profit are bulwarks that will not be entrenched on the world stage through the emotional whims of patriotism or tradition—moments after the same (or better) product is available at a lower cost from a different supplier, buyers will flock to the new supplier.

Rare earth metal mining operations and processing outside of China showed a shrinking bottom line as profit opportunities dried up when mining companies faced the option of selling at or near the prices ushered in by China or ceasing operations altogether.

China beat the United States because its miners and processors could sell at a lower cost due to economies of scale and a government willing to set low prices to gain market share. This drove suppliers in other parts of the world out of business through a decreasing profit margin, much the same way the arrival of a large chain store closes the door of "mom and pop" specialty shops. When profits drop due to a changing marketplace, decreased demand, increased upkeep costs, or a number of other factors, it is often cost-effective to cease mining operations at a location. It is not the place of these private corporations to mine any metal with the hopes of maintaining a healthy domestic supply and the best interests of the country if the monetary reward is not present.

Making use of rare earth metals in various incarnations allows for significant gains in efficiency and power—a neodymium magnet motor can outwork an iron-based magnet motor of more than twice its size—but these benefits are not without a substantial price. Rare earth magnet components often cost ten or more times the price of their less efficient, more common counterparts, and any disruption in supply will only lead to a widening of the price gap. When faced with a long-term drop in the supply of rare earth metals, manufacturers will be forced to choose between passing the costs onto the consumer and in the process risk losing market share, or selecting cheaper, older parts and manufacturing methods—the same ones many of the rare earth metals helped replace—that would lead to inferior products and eliminate a number of technological advances.

Product costs and creature comforts are not the only areas to falter during a supply disruption; entire movements could suffer when substitutions are made for vital rare earth metals. The rare earth metals dysprosium, neodymium, terbium, and lanthanum are the four-pronged linchpin of efforts to create an environmentally friendly transportation sector, with each metal needed in massive quantities if the electric car revolution succeeds in removing combustible engines from our roads and highways.

There are over thirty pounds of rare earth metals inside of each Toyota Prius that comes off a production line, with most of that mass split between rare earth components essential to motors and the rechargeable battery. Of this thirty, ten to fifteen pounds is lanthanum, with the lanthanum used as the metal component of nickel metal hydride (NiMH) batteries.[3] As the first generation of hybrid automobiles reaches the end of its lifetime, owners will be forced to replace their battery or move on to a different car, with both alternatives bringing an uptick in rare earth metal consumption.

The amount of rare earth metals needed to create of a state-of-the-art wind turbine dwarfs that needed for an electric car, with five hundred pounds of rare earth metals needed to outfit the motors and other interior components of a single energy-generating wind turbine.[4]

Looking for a way to end the green planet revolution? Cut off the rare earth supply line. Our future technology and way of life hinges on these metals, but before we extrapolate current political and economic situations to forecast the future, let's turn to the past. Humankind's relatively short working history with these metals began in the nineteenth century, as detail-oriented, hyper-analytical scientists who were one-part geologist, one-part chemist, and one-part physicist poured over ore samples from across the world to discern their origins.

CHAPTER 4

INSIDE A SINGLE ROCK

Each of the seventeen rare earth metals exhibits similar basic chemical and physical properties, with these similarities providing quite the challenge when it comes to separating them from one another in raw mineral ore. If you heat a mineral sample containing several of the rare earth metals to extremely high temperatures, it becomes difficult, if not impossible, to differentiate and physically separate each one because they share similar melting points. The rare earth elements are intricately bound to one another along with abundant elements like carbon and oxygen, making it impossible for industrious at-home refiners and large corporations to pick up a hundred pounds of raw mineral rocks and chip away for hours to separate the elements as one could do, in theory, with gold. Instead, concentrated acids and bases are needed to extract the individual elements, with chemists trying thousands of combinations before settling on the proper method to separate and purify a rare earth metal like cerium, a metal needed for use in pollution-eliminating catalytic converters, from a sample of bastnäsite or monazite.

The first of the seventeen rare earth metals discovered was yttrium, a discovery that began with the teamwork of a scientific "odd couple." Johan Gadolin, a thirty-one-year-old native of Finland who abandoned a career in mathematics to become a chemist, examined the composition of a black rock he received from Carl Axel Arrhenius, a lieutenant in the Swedish Army. Arrhenius fashioned himself as something of an inquisitive chemical mind himself and took the coal-like rock from a feldspar quarry he visited in Ytterby, Sweden. After months of investigation, Arrhenius bore little to show in the way of the chemical composition of the sample—a sample he dubbed "ytterbite" in the meantime—so he passed the rock on to more apt hands.[1]

"Apt" is an almost underwhelming way to describe Johan Gadolin. The bulk of Gadolin's graduate training centered on the in-depth analysis of iron ore samples, training that made the chemist perfect for answering questions about Arrhenius's specimen. Gadolin changed the way scientists passed down their knowledge to the next generation during his time as a professor at the Royal Academy of Turku. Gadolin is believed to be the first professor to create hands-on laboratory exercises for his pupils, the forerunner of the modern university laboratory courses that provide valuable instructional time for undergraduates and give graduate students an opportunity to sharpen their teaching skills.[2]

Teasing his way through the sample, Gadolin's analysis showed over a third of the ytterbite to be composed of a never-before-seen chemical compound, one he dubbed "yttria." Additional analysis in the following years revealed that Gadolin's yttria was not an element but rather a molecule featuring the elements yttrium and oxygen bound together. Glass is a great example of an oxide you come in contact with on a daily basis, a solid comprised of an atom of silicon bound to two atoms of oxygen. These "oxides" are prevalent in minerals, serving to both stabilize and preserve the metal in mineral samples and at the same time necessitate difficult steps to separate the desired metal from oxygen for use. Bonds between rare earth metals and oxygen to form solid compounds would plague scientists in their discovery of the individual metals over the next two centuries.

Although Gadolin did not break down the composition of Arrhenius's ytterbite into its most basic components, he is forgiven due to the extreme difficulty inherent in this task at the dawn of the nineteenth century. Had Gadolin made his discovery in 1842 instead of 1792, advances in the field no doubt would have nudged him to the discovery of elemental yttrium.

The hindsight of history allows us to note that Gadolin did miss the opportunity to make a second discovery—Arrhenius's ytterbite sample contained a second then-unknown element, beryllium, an element now deemed vital to US national security due to its inclusion in next-generation fighter jets and drones. Beryllium did not remain hidden for long, as scientists uncovered the abundant metal five years later.[3] To honor Gadolin's work, the rare earth metal gadolinium was named for him; it is used to create the memory-storage components of hard drives. And if you're a fan of resourceful middlemen, Carl Axel Arrhenius did not fare too poorly either, since the Swede is often given partial credit for the discovery of yttrium because of his initial effort to characterize the coal-like rock he dubbed "ytterbite" and for his decision to hand the sample over to the capable Gadolin.

GOOD COMPANY

For over three decades, one metal sat on the periodic table, playing the role of an impostor hiding in plain sight, fooling a man now viewed as one of the eminent chemists of the day. Carl Gustaf Mosander, a Swedish chemist who split his days and nights between teaching at Stockholm's Karolinska Institute and poring over the vast collection of mineral samples at the Swedish Museum of Natural History, discovered three elements, but he went to his grave believing he had discovered a fourth. Scientists emanating from Sweden and Finland during the eighteenth and nineteenth centuries made a disproportionate number of findings when compared to the rest of Europe and the world, in part because of the early movement to organize scientists in the region through the 1739 founding of the Royal Swedish Academy of Sciences, a group that leads the selection for the Nobel Prize in physics and chemistry to this day. The two countries, both allied under the Swedish flag at this time in history, also benefited from a comparatively strife-free nineteenth century while England, France, and Spain were continuously embroiled in war at home and abroad.

While examining a particularly prized metal oxide sample in 1841, Mosander's eye settled on an unusual substance never seen before: didymium. Carl Mosander found didymium within a metal residue he had been studying for many years. The sample fascinated Mosander to such an extent that he gave the specimen a special name, the *lantana*, a title derived from the Latin *latet*, meaning "hidden." The resources of the time period—a barrage of chemical reactions employed to separate a single substance from a mineral sample—agreed with Mosander's finding, and didymium soon joined the remaining known elements on the periodic table under the abbreviation "Di."

Shortly after Mosander's discovery, glass workers began adding the newfound metal to blacksmith goggles. The added didymium blocked the extremely intensive flashes of light created during smithing impacts. The flashes of light put a number of blacksmiths out of work because it led to a painful burning of the corneas known medically as photokeratitis. The United States also made use of didymium's proficiency for selectively passing through light, using the metal to create a method for invisible Morse code transmission in World War I. The transmitters used didymium-doped lenses to make the "dits" and "dahs" of Morse code invisible to the passerby unless they looked on with the correct viewing filter.[4]

Fifteen years after Mosander's death, other scientists pried into the fundamentals of didymium and suggested Mosander's discovery may not be an element but rather a combination of yet-to-be-isolated new elements. This line of reasoning held true, with Carl Auer von Welsbach, an astute rare earth metal scientist who would go on to improve Thomas Edison's model of the lightbulb, teasing out the rare earth elements praseodymium and neodymium—"green" didymium and "new" didymium—forty-four years after Mosander announced the discovery of didymium.

Like Gadolin before him, Mosander's oversight is no indication of his ability as a chemist. He discovered three elements—the rare earth metals lanthanum (eventually teased from the aforementioned lantana), erbium, and terbium—cementing him as one of the preeminent but sadly overlooked scientists of the modern era. If even one of the basic physical tests, spectroscopy, available to his successors just decades later, had been within his reach, Mosander would have easily separated praseodymium from neodymium and gone to his grave having discovered a total of five elements—all rare earth metals.

Despite the eventual separation into praseodymium and neodymium, the use of didymium continues to evolve. Oil refineries use the mixture of two elements as a catalyst in petroleum cracking, a heat-intensive process necessary to break down carbon to carbon bonds present in extremely large molecules en route to the culling of octane for use in gasoline.[5]

OLD-FASHIONED HI-TECH

You may be asking yourself, how would scientists in the pre-light-bulb-and-electrical-outlet era of society determine if they held in their hands a newly discovered element? By setting their hard-earned sample on fire. The spectroscope, a simple handheld device, allowed chemists, physicists, and geologists to light a sample of their newfound element ablaze and view a series of colored lines unique to each individual element. When one heats a metal until it glows, it gives off thermal radiation. One observes thermal radiation when using an old coat hanger to heat marshmallows over a campfire as the metal undergoes an orange to yellow to white color transition in the midst of the fire. This radiation is not the type that will harm an individual and doom him to a life of chemotherapy; it instead gives us insight into the constituents of the burning material when viewed through a spectroscope.

The spectroscope itself is a very basic instrument—a short tube with a prism inside—that allows a viewer to peer through one end and see bands of light reflected by the prism. Once scientists gathered the numbers of lines, spacing, and colors associated with known elements like iron and carbon, the spectroscope became the predominant tool for detection of new elements, since any substance giving off a heretofore unreported pattern could be a new element.

Each element viewed through a spectroscope gives off a specific set of colors and spacing when viewed, much like a person's unique set of fingerprints. My fingerprints are sufficiently different from yours, just as the color and positions of bands of light given off by oxygen are distinct from those arising from a sample of iron. Hydrogen gives off four visible lines in its spectral emission signature—from left to right a purple line followed by a blue line, a green-blue line, and a red line. When excited by heat or electricity, indium, a metal that sits two spots below aluminum on the periodic table, emits two lines—a light-blue line and a dark-blue line. Hydrogen and indium are simple examples, with the spectral line combinations increasing in complexity across the periodic table. For example, iron and phosphorous give off dozens of lines varying in color from red to violet.

As an added benefit, early spectroscopes were relatively inexpensive and small in size, making them the perfect tool for ambitious element seekers in the nineteenth century. As seekers sifted through crushed samples of rock from across the world, the quizzical scientists would heat pieces to very high temperatures and view the lines given off by thermal radiation through their spectroscope. If the lines observed differed from those recorded for the "known" elements, the researcher might be lucky enough to have found a new element. Combinations of elements, however, could give readings one might confuse for a newly discovered element, as well as samples not properly purified or cleaned.

Let's take our previous examples of hydrogen and indium. Separately, the elements give off four and two lines, respectively, when pure samples are viewed using a spectroscope. If hydrogen and indium are present together in a sample, the viewer spectrometer would see the combination of their spectral lines, for a total of six lines. It would be quite easy for a scientist, mental acuity blurred by the possibility of discovering a new element and entering into chemical immortality, to incorrectly determine a complex spectra set containing seventy to eighty lines belonging to a new element and not a combination of existing known elements.

How did cunning scientists eliminate an embarrassing false detection problem? They did not. While the scientists purified their samples as much as possible—often going to obsessive-compulsive lengths in doing so—purifying a simple compound containing the target element in the belief that the element itself had been isolated often occurred. It would not be until a new generation of scientists, using an increased knowledge base and advanced technology, examined their forefathers' results and found them lacking.

CHAPTER 5

A SCIENTIFIC COLD WAR

During the 1950s and 1960s, the United States and the USSR dedicated resources to discovering new elements, particularly metals. US scientists at Lawrence Berkeley National Laboratory and Soviet researchers at the Flerov Laboratory of Nuclear Reactions along with enormous amounts of money were dedicated to these projects in the hope of finding a substance similar to uranium that could be manipulated to build weapons of mass destruction.

When is the discovery of a new metal viewed by national governments on par with a successful military weapon launch? Such victories are possible, and necessary, when the two superpowers are locked in a multi-decade struggle in which the simplest of physical battles would prompt the launching of hundreds of nuclear missiles across the globe, quite possibly ending life on Planet Earth.

Even though many of these rare metals do not have any commercial application whatsoever (at least for now), they can make an amazing impact on the geopolitical climate. A Cold War race between the Soviet Union and the United States led to the discovery of a number of new elements in the hope that one or more of them could be used in lieu of uranium or plutonium. Almost all of these newly discovered elements are classified as metals—metals with names conjuring historical people and places like americium, berkelium, and curium. While these metals did not lead to the creation of a Cold War–era weapon of mass destruction, this bizarre form of "test-tube combat" fought for nation-states by their resident scientists allowed resource-plenty nations to wage a bloodless war in laboratories and stave off tremendous loss of life.

THE FLEETING EXISTENCE OF SYNTHETIC METALS

Modern scientists continue to search for new elements, with most existing for only a few seconds within particle accelerators. Due to the progression of the periodic table, all but one of the elements—all synthetic—discovered since the early 1980s—hassium, meitnerium, roentgenium, copernicium, flerovium, livermorium, and the five generically named "unun-" elements—are metals or are expected to exhibit metallic characteristics if scientists are ever able to create a sufficient number of atoms to test.

The synthetic elements form what could be called the "kiddie table" of chemistry. Much like newborns, the creation of a new synthetic element is big news, and the synthetic elements owe their existence to the hard work of their chemical parents. This special group is not very useful, not at the moment, but the amount of effort that goes into learning about the synthetics is extraordinary when compared to the rest of the periodic table. Whispers fill the air about the "next" synthetic elements to be found, with educated guesses about their possible physical properties made based on the elements that sit directly above their open space on the periodic table. Of the one hundred and eighteen elements currently included on the periodic table (a number that changes frequently, with elements removed or added and placeholders put in position every couple of years), nineteen do not exist in any form on our planet. No matter how many tons of rubble you sift through, how deep in the oceans you dive, how bizarre of a locale, you will not find a single atom comprising 16 percent of the periodic table. These elements are products of scientists and engineers slamming atoms and particles together at speeds of thousands of feet per second. Why would scientists expend time, knowledge, and an immense amount of energy to smash atoms? They did so in the hope of sifting through the aftermath for a lucky strike, even if the new synthetic element created by the strike exists only for a fraction of a second before disassembling at a particle level and becoming a more mundane but stable element like lead. Through all the joy of success and the monotony of failure, researchers performing these experiments are largely forced to keep their work wrapped in a veil of secrecy. The element of secrecy was not in place for the normal academic reasons—to prevent other scientists from trumping your work—but for the survival of your family, friends, and country.

Prior to the rise of a twenty-four-hour news cycle, maintaining scientific secrets under the veil of national security lacked the level of difficulty it does now, but information and facts still slipped through the cracks through unusual channels. An unlikely culprit, the comic book hero Superman, is to blame for spreading the details about a piece of scientific instrumentation known as the cyclotron. Ernest Lawrence, the father of nuclear fission, created the first cyclotron at the University of California–Berkley in 1932, giving the United States more than a decade head start over the Axis powers. Germany would not construct a cyclotron until the last months of World War II.

The cyclotron is what an action hero might call an "atom smasher"; the device is an early type of particle accelerator that shoots ionized particles along a spiral path toward a monitored target at enormous speeds. Scientists using cyclotrons hope to capture any altered particles arising from collision for further study. The cyclotron is responsible for the most important elemental discoveries of the twentieth century: the 1940 discovery that plutonium and neptunium are created when neutron-heavy forms of hydrogen traveling at near-relativistic speeds impact a uranium target. The United States kept the creation of plutonium a thinly veiled secret and in the five years after its discovery built and funded the Manhattan Project in order to scale up production of plutonium and create the atomic bombs that would fall on Hiroshima and Nagasaki in the summer of 1945.

When the comics section of the morning paper featured a scientist discussing the use of the cyclotron in a daily installment of the *Superman* strip, the eyes of those within the United States Department of War (the precursor of the US Department of Defense) widened. Alvin Schwartz, the writer of the *Superman* newspaper dailies in April of 1945, placed the cyclotron in the story as part of a villainous professor's attempt to test the limits of Superman's invulnerability by zapping the last son of Krypton with three million volts of electricity. It was not at all out of character for superheroes to take on military-minded opponents during World War II, with Superman regularly fighting Nazis and becoming the subject of real-life attacks by the National Socialist Party when the official newspaper of the Schutzstaffel paramilitary group (better known in the eyes of history as the villainous "SS"), *Das Schwarze Korps*, published an article condemning one of the Jewish cocreators of Superman, Jerry Siegel, and a two-page comic printed in the February 27, 1940, edition of *Look* that detailed how Superman would quickly end World War II.[1]

The insertion proved to be a clever use of a modern marvel of scientific achievement, but Schwartz nevertheless received notice from the long arms of the US government. Spreading information about nuclear physics and its application, however ridiculous, was seen as straying from the best interests of the American people. After an unsuccessful attempt to stop the strip from reaching hundreds of thousands of homes across America, National Periodicals (now the Warner Brothers–owned DC Comics) agreed to remove future references to atomic energy and weapons from the newspaper strip and line of comic books.[2]

This ban of sorts would turn out to be a failure: nine months later the January 1946 issue of National Periodicals' *Superman*, just the thirty-eighth issue of the series, featured a story titled "Battle of the Atoms," a tale wherein archvillain Lex Luthor detonated an atomic bomb—one different than the bomb created by the efforts of the Manhattan Project but close enough to raise the ire of the Department of War once more—in a futile attempt to defeat Superman.

THE ELEMENT OF WAR

If one rare and elusive synthetic metal is to be selected as a singular key to the twentieth century, plutonium would easily be the choice. Although we now know that it exists naturally in extremely small quantities spread like ashes over the planet, the first meeting between man and plutonium came in synthetic form. Glenn Seaborg, who would later have an element named in his honor, brought plutonium into the modern purview through a series of experiments carried out by his research team at the University of California–Berkeley at the dawn of World War II. He reported his successful experiments, in which he bombarded a sample of uranium-238 with atoms of deuterium, a form of hydrogen that carries with it an extra neutron in 1940, changing the world. Seaborg was only twenty-eight at the time, an age at which most individuals are still working to earn their PhDs. These experiments created a scant amount of plutonium, with the US government cheering Seaborg on in this effort, and soon the government brought him on as a key player in the Manhattan Project.

While the supplies of plutonium used in nuclear weapons and in academic research are human-made, plutonium is present in extremely small quantities in the environment. Scientists believed plutonium was a wholly synthetic element, placing the metal in the realm of einsteinium and fermium for a short time. A 1971 Los Alamos National Laboratory study found a naturally occurring form of plutonium in California strata dating to the Precambrian Age, with this trace likely billions of years old and having its origins in the aftermath of the big bang and the accompanying celestial events that led to the creation of Planet Earth.[3] Current estimates place the world census of naturally occurring plutonium at a whopping one-twentieth of a gram.[4]

Plutonium, as you likely know from its destructive applications, is highly energetic. If you held a piece of the silvery-white metal in your hands, the cube would gently warm your palm within a matter of seconds, thanks to the atom's steady release of alpha radiation. If given the chance, please don't ever hold a piece of plutonium—if you do, it will be one of the last items you hold before checking into an isolated room with a hospital bed. The Greek nomenclature used to describe radioactive energy does not help one who is beginning to learn the details of the phenomenon, and these details continue to puzzle college and graduate students attempting to keep the types of radiation organized in their heads among a blizzard of facts and pathways needed to navigate a day's work. Alpha particles, gamma rays, beta particles, and the like are all shorthand for the type of energy and physical material we know to be emitted through radioactive processes. Each alpha "particle" is made up of two different particles: two protons and two neutrons, making an alpha particle an atom of helium that has had its electrons stripped off. A beta particle is a lot simpler to understand, since it is the same as an electron. A gamma ray, other than being the form of radiation responsible for giving the world the Incredible Hulk, is, like the name suggests, not a physical particle but a form of energy with a very short wavelength. The short wavelength allows the energy to penetrate deep into substances it comes in contact with, making the gamma ray a very powerful and quite dangerous form of radiation.

While plutonium is powerful, not every atom of plutonium is created equal. The plutonium-239 isotope is prized, while plutonium-240 and plutonium-241, each one becoming slightly heavier as an atom captures additional neutrons, are seen as impurities in many applications because they perform differently. The amount of these two isotopic "impurities" leads to the designation of grades for refined plutonium, dividing up samples into weapons grade, nuclear reactor grade, and so on. Weapons-grade plutonium is the purest, containing around 95 percent plutonium-239, while reactor-grade plutonium needs only half the amount of plutonium-239. From an energy-generating perspective, the other isotopes of plutonium are viewed as less energetic dead ends.[5]

BORN OUT OF A NUCLEAR EXPLOSION

Not all elemental discovery expeditions need fancy cyclotrons or rare metal targets to be successful. Due to the incomplete nature of the periodic table as little as seventy years ago, all one needed was a keen eye and a nuclear bomb. The first synthetic elements created outside the walls of a laboratory, a pair of twins, made themselves known in the aftermath of the first hydrogen bomb test. On a calm eighty-degree morning in November of 1952, scientists let loose the secret operation "Ivy Mike" along a tiny string of islands on the Enewetak Atoll (islands now part of Micronesia's Republic of the Marshall Islands). The high-power atomic detonation created a crater one mile in diameter and half a football field deep along with elements einsteinium and fermium. The elements were named for two living legends of science walking the earth at the time: the well-known and enigmatic Albert Einstein and Enrico Fermi, an Italian scientist (and hero of many academics) responsible for turning a tiny space under the University of Chicago's football field into the world's first nuclear reactor. Fermi had a personal connection to the persecution felt by many leading up to World War II, as the Nobel Prize–winner left his home country of Italy to protect his wife from anti-Jewish legislation. US scientists sifted through filters flown through the debris clouds created by this hydrogen bomb, looking for anything unusual in the aftermath of this historic event that sounded the silent opening shot of the scientific leg of the Cold War.

A handful of uranium atoms at work during the Ivy Mike test took on far more than the usual amount of neutrons expected in a nuclear detonation. The energy and neutrons released in the course of the twelve-kiloton explosion—four times the combined might of the atomic bombs dropped on Hiroshima and Nagasaki in the final days of World War II—provided the proper environment for a nuclear "primordial soup," leading to creation of two never-before-observed metal elements.

The creation of einsteinium and fermium did not mark the first time scientists brought a new element into existence. Scientists battled each other and Mother Nature in the search for technetium during the 1920s and 1930s, a metal first suggested to exist eight decades prior, with the mantle of discovery heavily influenced by political ties.

POLITICS AND DISCOVERY

While Ida Tacke Noddack excelled at the meeting place of geology and chemistry—a very useful skill set in the nineteenth and early twentieth century as scientists raced to fill holes in the periodic table to gain academic superiority—she faced her most complex problems outside the walls of a chemistry lab. The German-born chemist discovered the forty-third element in 1925, giving it the name masurium in honor of a region in Poland where two grisly battles between Germany and Russia took place during World War I. The First World War marked a turning point in the intermingling of science and the battlefield. France and Germany made liberal use of nonlethal irritants in warfare like tear gas in the early days of the conflict. By the end of 1915, Germany moved on to far more deadly implements of chemical warfare. By the end of World War I, soldiers deployed chlorine, phosphine, and mustard gas on battlefields, killing tens of thousands.[6]

In hopes of gaining the upper hand over an enemy, new elements discovered in the decades prior to World War I became targets for military research, with scientists all over the world obliging. Scientists overlooked scientific achievements concerned with the common good of humankind in lieu of studying the properties of newly found elements in the expectation of turning any of the elements into a weapon of war. Given the historical setting, a German scientist calling an element masurium is not too far removed from a World War II–era Japanese scientist christening a new discovery with a name associated with his country's victory at Pearl Harbor. History paints Noddack's selection of the name with a narrow brush, often suggesting that she chose the name to be symbolic of these German victories against outside powers. In reality, Noddack did not select the name—her coworker and soon-to-be husband, Walter Noddack, chose it. Regardless of the origin, lending the name masurium to the discovery did not help Noddack's efforts to prove to the scientific community that they indeed found the element, in a time where fighting to gain worldwide acceptance for one's scientific claim often proved more difficult than the research itself.

Chemists and physicists endlessly critiqued the method by which Ida Noddack detected the presence of masurium because her claim centered on a very weak signal obtained when bombarding a sample of the mineral columbite with electrons, a signal so small it is easy to write off as background noise. Independent researchers experienced difficulty in reproducing Ida and Walter Noddack's experimental results en route to claiming the discovery of element forty-three. Ida Noddack's gender, unfortunately, diminished the efforts of her research in the eyes of suspicious scientists, a problematic child born out of the "boys club" of nineteenth- and twentieth-century science. Lise Meitner, a key member of the Nobel Prize–recognized team that discovered nuclear fission and one of the few women of the era with enough power to sway opinion on her efforts, did not help Noddack's case, calling her a "disagreeable thing" in a letter to her Nobel Prize–winning colleague Otto Hahn.[7]

If one looks for the location of masurium on the periodic table, it will not be found, not even after surveying the quarantined rows huddled at the bottom. Twelve years after Noddack reported the finding of masurium, credit for the discovery of element forty-three had been transferred to a pair of Italian scientists, Carlo Perrier and Emilio Segrè. The duo successfully detected and isolated element forty-three by performing experiments on radioactive discards from a cyclotron decommissioned from the US Lawrence Berkeley National Laboratory. Perrier and Segrè studied a piece of foil created from element forty-two, the metal molybdenum, which researchers bombarded with neutrons in the process of cyclotron use, hoping to find atoms of molybdenum transmuted into a new element. Success came quickly, with the pair naming this new element technetium.[8]

Despite substantial controversy throughout Noddack's career, hindsight gives her the last laugh. Ida Tacke Noddack is now noted for early theories suggesting the possibility of nuclear fission—a phenomenon by which heavy elements split into smaller fragments. Her suggestion went overlooked in the years before Meitner and Hahn's momentous findings on the subject due to public fallout from the masurium debacle. Noddack also codiscovered rhenium and received a trio of nominations for the Nobel Prize in chemistry while still in her late thirties and early forties.

When Perrier and Segrè announced their discovery of the forty-third element in 1937—the same one Ida Noddack reportedly found—the name technetium stuck, but the metal would go without a major use until its introduction into nuclear medicine as a radioactive tracer. Oddly enough, this would be the third name for the element. Dmitri Mendeleev, the father of the modern periodic table, predicted the existence of an element in the same position where technetium would eventually fit, along with a number of correct properties more than fifty years before Noddack claimed to have discovered masurium. Mendeleev bestowed upon the hypothetical element the name *ekamanganese*, digging deep into historical languages to use the Sanskrit prefix "eka," which is commonly translated as the English word "identical." Mendeleev based his predictions for the properties of ekamanganese on those known for manganese, the element that resides above it in the periodic table. Ekamanganese was not the only element predicted to exist by Mendeleev; the Russian chemist successfully theorized the existence and characteristics of scandium, gallium, and germanium, as well as their eventual position in the modern periodic tables.[9]

WHY RESEARCH WHEN YOU COULD BE BUILDING BOMBS?

So why did the United States and the Soviet Union dedicate a legion of their finest minds and a nearly unlimited budget to search for new elements? If a suitable atomic weapon could be made from a new element—one more powerful than the plutonium and uranium cores at the heart of the era's nuclear weapons—the lucky country would automatically become the biggest bully on the world's block. Such a weapon would send an opposing country into a defensive position until its scientists could replicate or improve the weapon. From another perspective, the element would not necessarily allow for the creation of a more powerful weapon; it would be a major advantage for either side if the chemical core of the weapon was cheaper, allowing for increases in production and deployment across the globe.

As the Cold War dragged on, it became apparent that it would not look like other wars due to the looming threat of mutually assured destruction. Instead, the war became a struggle over perceived abilities, know-how, and theoretical strength, with the Soviet Union and the United States equipping small armies of scientific geniuses with state-of-the-art laboratories and tasking these strike forces to create new elements in the name of national security. Meting out scientific mettle with cadres of strategically positioned scientists boiled this field of research down to yet another theater of the Cold War as early as the 1950s.

As years of passive-aggressive conflict went by, the goal evolved away from national security toward one of national pride. If a newly observed element failed to become the basis of a world-changing superweapon, all was not lost, as its discovery was a victory in and of itself, another point on a perverse elemental scorecard kept by the Soviet Union and the United States. The race became a barometer for exactly how adept the physicists and the chemists of each country were in their fields and in the public consciousness.

For example, if the Soviet Union could confidently dedicate some of its best minds to bring into existence a new element that is with us for only a fraction of a millisecond, what else are they capable of creating in clandestine projects carried out behind closed doors? Thoughts just like this one sent the minds of state officials and generals, along with ordinary citizens and schoolchildren of both countries, racing. What clandestine advancements are being made on the other side?

The discovery of an element often came along with the spoils of naming the new finding, another feather in the cap of statesmen who sought to christen this new entry in the history books with a name invoking one of the heroes of science in their home country (like fermium or lawrencium, both named for champions of nuclear physics within the United States), or to name it for the place of discovery or the country itself (element 105, dubnium, is a great example of this manner of naming; the element is named in honor of the Soviet Union's premier government laboratory).

The scientists given the opportunity to join this search were more than happy to loan their talents because they gained access to a wealth of funding and academic power along with a chance at immortality, all while serving the perceived greater good of their country against life-and-death stakes. Discovering an element is the scientific calling card of a lifetime, often paving the way for deification in university and government laboratories and for Nobel Prize consideration every year for the rest of the researcher's life.

Though the race between the Soviet Union and the United

States to discover new elements died out decades ago, a new one with a similar context could be brewing. A myriad of weapons devices used by the United States and a handful of other countries rely on rare earth metals to operate. Neodymium and its neighbor on the periodic table, samarium, are relied on to manufacture critical components of smart bombs and precision-guided missiles, ytterbium, terbium, and europium are used to create lasers that seek out mines on land and under water, and other rare earth elements are needed to build the motors and actuators used for Predator drones and various electronics like jamming devices.[10]

The problem for the United States is obvious: China, a country with radically different political beliefs that is often painted as—at best—a reluctant ally of the United States and at worst a future conqueror, controls all but 3 to 5 percent of the rare earth metals market. Would our government want to go to someone it typically dislikes to purchase anything it relies on, let alone a material the United States is stockpiling and could one day use against them? This is the exact situation the United States is in—if a country wants the supply of rare earth metals to grow and advance their stockpile of certain hi-tech weapons and battlefield resources, that country is forced to deal with the corporations of China and its government, entities with little incentive other than monetary ones to make a deal. A dire situation such as this one is sending countries down bizarre avenues to become more self-reliant, including the pursuit of rare metals in the belly of nuclear reactors.

CHAPTER 6

CREATED IN A NUCLEAR REACTOR

Several rare metals absorb phenomenal amounts of neutrons, making the metals perfect for use as control rods in nuclear power plants. At the core of nuclear fission reactions within these plants, elements are also created, with several rare metals, including the medically applicable rhodium, synthesized in these reactions and gathered for sale and industrial use.

Currently, there are thirty-eight naturally radioactive elements known to humankind. Most of us know the names of the big boys on the block—uranium and plutonium—but past that, our face-value knowledge takes a steep drop. This is understandable since we (hopefully) do not come in contact with any of them in our day-to-day lives. The first synthetically created element, technetium, is radioactive, as is one of the rare earth metals, promethium. In our lifetime, we may come in contact with technetium because the metal's radioactive properties and short half-life are leveraged in the use of technetium as a radioactive tracer in humans to give physicians and surgeons diagnostic insights prior to an invasive procedure.

Each element from position eighty-four to the end of the periodic table at one hundred and eighteen is radioactive, and of these thirty-six elements, only twelve are available in large enough quantities to be useful to humans. This scarcity is changing because an unlikely source—cast-aside nuclear fuel rods—could become a key for elements like curium and californium: naturally radioactive metals that are useful in fields outside of basic scientific research once pried from their uranium oxide fuel rod birthplace.

The prized and widely used metals rhodium, ruthenium, and palladium are also created in significant quantities within fuel rods. Fuel rods are roughly the same length as a fishing pole, and the rods spend their working lives in the same manner, poking through a pool of water. Instead of probing lakes for fish, the rods use their body of water for cooling and control of the millions upon millions of individual reactions occurring within each passing moment.

Deep in the interior of nuclear power plants the fuel rods are arranged in arrays within a cooling pool to maximize safety. The goal is to allow the heat generated from the billions of neutron additions to safely flow through the water—without the liquid, the heat created as a result of reactions ongoing within fuel rods would quickly overrun any containment units and lead to a meltdown. Water is chosen as the mediating material due to its ability to take on a substantial quantity of heat before evaporating. Engineers within nuclear power plants alter the pressure in fuel rod containment facilities—a major safety issue for those working in and around the rooms but one that allows the water to mitigate much more heat than normal at atmospheric pressure before boiling off.

Thousands of oblong pellets of uranium oxide salt are arranged in each fuel rod, akin to a bevy of extremely energetic Easter eggs packed into a twelve-foot-long metal basket. After initiating a nuclear fission reaction by slamming neutrons into atoms of uranium, energy is given off as heat. The release of heat is funneled on to generate a steady stream of steam used to spin arrays of turbines attached to generators, leaving us with the electricity used to power homes and businesses across the world. France is the world's leader in widespread use of nuclear energy, providing over three-quarters of its electricity through nuclear power in the first decade of the twenty-first century, along with generating a significant portion of that energy, 17 percent, through the use of recycled nuclear fuel.[1]

Once the fuel rods are placed in the proper position in a nuclear power plant, an initiating event is scheduled to kick-start the reactions in a fuel rod that will lead to the downstream production of nuclear energy. For new nuclear plants, placing a previously used fuel rod in the reactor is enough to initiate a safe and stable chain reaction, but if no used fuel rods are at the disposal of the facility's engineers, mixing the alpha particle emitter polonium with beryllium will lead to the release of enough neutrons from the beryllium sample to start reactions in the fuel rods.[2] The polonium-beryllium initiation method is extremely powerful; it was also used to catalyze detonation of early nuclear weapons designed around a spherical plutonium or uranium "pit."[3]

In time, the overall efficiency of energy-creating reactions within a rod decreases. When the fuel rod is deemed no longer sufficient to carry out nuclear reactions in a reactor, the uranium is designated as "spent." The atoms inside the fuel rod are still capable of undergoing a nuclear reaction, but the physical distribution of atoms in the rod is much different than when its time of service began. Metals and molecules have migrated from their starting points in the rod in search of localized hot and cold spots, and a symphony of metal atoms nonexistent at the time of installation, including palladium and curium, have been created as uranium atoms took on neutrons during energy-creating nuclear fission reactions.

The next phase of life for a spent fuel rod is a controversial one. The rods are either designated for containment in a lead-lined waste facility in a remote location, or, in an increasingly common development, a series of reprocessing steps are undertaken to recover the uranium inside for use in applications where the standard of efficiency is less stringent than energy production. It is here, in the fuel rod–reprocessing step, that a variety of desirable rare metals and scarce radioactive elements with useful applications created during years of service are gathered for use in other venues.

Technetium is made in small quantities, while the three metals squeezed between technetium and silver on the periodic table—ruthenium, rhodium, and palladium—are created by neutron collisions through the course of nuclear power production. Over the years of a fuel rod's service life, the atom-by-atom conversion leads to the buildup of tangible quantities of these metals, so much so that scientists have determined methods by which these elemental metals can be recovered from used radioactive fuel.

When the spent fuel rods are divided into smaller sections and combined with a strong acid, the uranium pellets dissolve. While uranium from the fuel rod readily dissolves, the rhodium and other metals created through two to three years of energy production will not. Instead, these metals form small clumps of particles in a fashion akin to tiny nuggets of gold deposited in a riverbed. The particles are separated and further concentrated with additional steps until nearly pure samples of individual metals are present.

Despite the quantities of new metals to be gained, the practice of nuclear reprocessing is frequently under scrutiny. Profits attached to the practice are eaten away by costs associated with labor- and time-intensive aspects of reprocessing. Uranium fuel poses an ever-present danger during the reprocessing period since, once uranium and plutonium are separated from their metal housings and dissolved in acid, it is still theoretically possible (although extremely unlikely) for them to gather in localized hot spots within the processing tanks and reach dangerous critical mass. Even if the economic hurdles and safety issues are overcome, the inherent nature of reprocessing sites and the substantial quantity of nuclear fuel within their walls could leave them vulnerable to direct attacks from terrorist groups or the theft of still-fissionable nuclear material.

It would be foolish to think an attack making use of nuclear material en route for reprocessing would not be devastating. Even if the attackers failed to turn stolen spent fuel into a high-power nuclear weapon, threats will forever loom from less scientifically advanced attacks stemming from the addition of radioactive waste into an existing explosive device or a strike on a nuclear reprocessing facility that would turn the entire site into an unconventional dirty bomb. Such an attack could exact minimal physical damage and still render the surrounding area unfit for habitation for many years. The psychological toll would be unlike any disaster seen in the Western Hemisphere, with hundreds of billions of dollars necessary to decontaminate and clean the area and tremendous upheaval as several generations would find their lives and homes severely impacted in a single attack.

These fears are not merely the creation of a post-9/11 think tank but are a hypothetical plague that has occupied the highest office in the land for six decades. Presidents Gerald Ford and Jimmy Carter halted reprocessing of plutonium and spent nuclear fuel during their terms in office in an effort to stop the spread of national nuclear weapons programs and clandestine attempts to secure a nuclear device across the globe—a fear bolstered by ongoing tensions in India and Iran during the late 1970s.[4]

President Ronald Reagan lifted this ban during his tenure, only to have his successor, George H. W. Bush, prevent New York's Long Island Power Authority from teaming with the French government–owned corporation Cogema (now a part of the nuclear energy conglomerate Areva NC) to process reactor fuel. President William J. Clinton followed Bush's lead, while President George W. Bush went on to embrace nuclear reprocessing by forming the sixteen-country Global Nuclear Energy Partnership and encouraging private corporations to develop new reprocessing technology.[5]

This trend of "stop-start" policy on the matter reversed once again with President Barack Obama, who signaled what appears to be the death knell for commercial nuclear processing in the United States, at least for the first half of the twenty-first century. Fiscal concerns informed his decision to cancel plans to build a large-scale nuclear reprocessing facility in 2009 and a South Carolina reprocessing site in 2014.[6] At the moment, the United States does not reprocess reactor fuel previously used to generate power for public consumption; it instead chooses to focus recycling efforts on radioactive materials created in the course of scientific research.

Regardless of one's personal political views, the reticence of five presidents to pursue nuclear processing—Ford, Carter, G. H. W. Bush, Clinton, and Obama—should be a sign to those championing the cause of nuclear processing. Financial issues aside, concentrating large amounts of nuclear material in one area, no matter how secure, with hundreds, if not thousands, of workers coming in contact with the material makes the site ripe for thievery and attack. Acquisition of radioactive material by clandestine individuals is not isolated to action movie plots and Tom Clancy novels but is a plausible threat. A dirty bomb has yet to be detonated anywhere in the world, thankfully confining these radiological weapons to movies and novels, wherein the bombs play the role of an all-too abundant plot device and source of melodrama. The most feeble of dirty bombs needs only a sufficient source of radioactive waste and an explosive device to disperse the waste in order to render a location unfit for years. The psychological impact of the aftermath of a dirty bomb detonation at any site with a nostalgic bend—a Major League Baseball stadium or national monument, for example—would be a constant reminder of existing cracks in the seams of national security and a calling card for the culprits.

The combination of extreme destructive capacity, theoretically discrete size, and multiple decades of cleanup work before the city is habitable led Representative Peter King, chairman of the House Committee on Homeland Security's Subcommittee on Counterterrorism and Intelligence, to claim that Washington's "nightmare scenario" would be the detonation of a dirty bomb in New York City by a terrorist group.[7] This claim came on the heels of President Obama's 2014 declaration that he personally fears the detonation of a dirty bomb in Manhattan more than the impending threat posed by a revitalized Russian Federation. This fear may very well end the reclamation of rare metals created within fuel rods before the process gets on its feet.

Reprocessing plants pose a final, passive threat that may be a key element in why governments are not as aggressive as they could be in reusing radioactive fuels and, instead, choose to lock the waste away for future generations. Almost every step of a reprocessing effort creates additional radioactive waste. Liters upon liters of strong acids and harsh carcinogenic solvents are used en route to reclaiming metallic uranium and plutonium that can used in a new way. This "new" waste created in the dissolving states contains only a fraction of the radioactivity in a sample of reactor-grade uranium, but nevertheless, the radioactive waste must be locked away until the natural decay of radiation over time occurs. By recycling used reactor fuel, we are able to make additional use of these rare metals; however, in the process it is possible to create considerable quantities waste. This balancing act between the continued downstream use of reprocessed nuclear fuels and their metal products juxtaposed with the disposal of newly created waste will determine the future utility of nuclear reprocessing, with the fears of lost radioactive material entering the hands of malicious groups forever clouding the entire procedure.

Until a government or corporation discovers a method of safely mitigating radioactive waste short of ejecting spent fuel rods into space (no nations are considering such a folly at the moment, although I would wager the idea has been kicked around in jest), our leaders will be forced to choose between reprocessing and reuse of nuclear fuel and locking up the waste for a future generation to contend with.

ONE ELEMENT, DIFFERENT COMPOSITIONS

Each metal synthesized in the belly of a nuclear reactor exhibits a substantial degree of variety. Depending on the element, the atoms present can have a number of isotopic states—curium, for example, has twenty-one. Isotopic states of the same element are called isotopes, a word we take from Greek, meaning "equal place." Each isotope lays stake to an identical number of protons (without sharing an identical number, the isotopes would no longer be the same element), but where the isotopes differ from one another is in the number of neutrons present, the non-charged particles coexisting with protons at the core of an atom. It is common for a single isotope to predominate. Fifteen different isotopes of carbon exist, but the overwhelming majority of carbon in the soil, oceans, and our body is carbon-12, with this isotope making up 989 out of every 1,000 atoms of carbon. Most people know of another isotope of carbon, carbon-14. Carbon-14 accounts for slightly more than one atom in one trillion atoms of carbon. Carbon-14, in time, will eject the additional neutron it contains in its nucleus and become a run-of-the-mill atom of carbon-12—a phenomenal gulf is scarcity, but one scientists routinely use to their advantage. Carbon-14 decays at an extremely steady pace, with scientists taking advantage of this natural phenomenon to date biological materials with uncanny accuracy. The number of carbon-14 atoms in a sample is cut in half every 5,730 years, with this phenomena defined by scientists as the half-life of carbon-14. Living biological organisms increase the amount of carbon-14 in their bodies as they breathe, but the deceased are no longer warranted such a luxury. With the carbon-14 atoms stable after death, scientists can calculate the finite number of carbon-14 atoms in a sample taken from a long-dead plant or animal and discern its age with a reasonable amount of accuracy up to forty-thousand years.

Many of the metals born out of fuel rods are like carbon-14 in that they decay to a more stable form over time at rates long established by scientists. There is one substantial problem to be wrestled with when it comes to these newly synthesized metals that does not plague carbon: the freshly created isotopes of curium, americium, and the like are often radioactive. Their radioactive nature poses a health hazard to humans and a contamination risk to any other metals they come in contact with, as contact could lead to the irradiation of a previously nonradioactive sample of metal. These safety and handling issues must be addressed before these fruits of reprocessed nuclear fuel can be used for further energy production or applications in medicine.

Thankfully the half-lives of a handful of isotopes created are relatively short—the major isotopes of palladium, ruthenium, and rhodium exist only for a handful of seconds to a few years before settling down to stable forms—otherwise the safe use of these metals within our lifetime would be impossible.

The safer choice, at least from the perspective of a politician seeking to reduce spending and ensure that his constituents elect him to another term, is to forgo expensive waste-elimination costs and bury spent fuel rods in desolate areas for future generations to contend with. The latter option is a popular one. Fourteen deep geological repositories are spread across the globe, while another eight are in the construction phase in Finland, Japan, Canada, Sweden, and Germany. Deep geological repositories make use of preexisting rock and salt formations and the massive caves and kilometer-long boreholes dotting the interior to store the most dangerous of nuclear waste, filling in the gaps between metal caskets and their radioactive cargo with cement, crushed salt, and soil to isolate the waste and prevent future movement. With nowhere else to go, these primitive repositories serve to place a physical impasse between radioactive material and human life, while preventing any leaks into the groundwater supply.[8]

There is but a single active geological repository in the United States, the Waste Isolation Pilot Plant roughly twenty-five miles outside of Carlsbad, New Mexico. This is the United States' second deep geological repository—Nevada's Yucca Mountain Nuclear Waste Repository served a similar use for research purposes, but operations ceased in 2008. The site opened its doors in 1999 for the disposal of plutonium and other transuranic elements. The Waste Isolation Pilot Plant boasts a substantial physical barrier, a two-thousand-foot-thick salt bed formed two hundred and fifty million years ago. Two thousand feet underneath New Mexico's State Road 128 sits fifty-plus football-field-sized rooms intended to store waste for tens of thousands of years. Projecting the existence of any edifice tens of thousands of years into the future is a fool's errand. The minds behind the construction and ongoing use of the Waste Isolation Pilot Plant do not believe so, going so far as to hypothesize and create wear-resistant warnings to cater to humans in the future who venture near or possibly drill into the facility.[9] These underground repositories are not airtight vaults, either, as evidenced by the New Mexico Waste Isolation Pilot Plant closing for an extended period in 2014 after twenty-one workers were exposed to a radiation source due an unknown leak.[10]

PLACING THE CREATED METALS INTO CIRCULATION

Any metals recovered for commercial use (including the traditionally nonradioactive palladium, ruthenium, and rhodium) would need to be sequestered for decades and stored in high-cost facilities before they could be placed into circulation. Why? To give the metals time to decay to an acceptable level of radioactivity. Reaching a decay state with zero radioactive signal is not practical due to the enormous time spans necessary for near-complete radioactive decay. Instead, the level at which these metals are released into the wild will likely be the moment the radioactive threat is so low that the owners are not held liable for health impacts. After seven half-lives pass, slightly more than 99 percent of the original radioactive material has decayed. For ruthenium and rhodium accumulated as synthesis products from countless nuclear fission reactions en route to producing electricity, seven half-lives is not an insurmountable barrier to later use. Rhodium separated from fuel rods is extremely stable, with minuscule quantities of rhodium-106 present. The half-life of the rhodium-106 isotope is a mere thirty seconds, but such a short life is an exception to the rule. The radioactive isotopes of ruthenium created in nuclear fission reactions, ruthenium-103 and ruthenium-106, exhibit half-lives of thirty-nine days and three hundred and sixty-eight days, respectively. These numbers place countdown to safe use at a reasonable nine months for ruthenium-103 and seven years for ruthenium-106.[11] Ruthenium-106 is already used in medical research, as the steady stream of low-power energy emitted by the radioactive isotope can be used to experimentally treat certain forms of cancer, including malignant uveal melanoma.[12]

We are not blessed with the advantage of time when it comes to the isotopic forms of palladium present in nuclear fuel remnants. Approximately 17 percent of salvaged palladium is present as palladium-107, which has a phenomenally long half-life of six and a half million years. We do not need to use the rule of seven half-lives to see that waiting for palladium-107 to decay to a passable radioactive is out of the question. The long half-life and steady stream of radiation will prevent recovered palladium from serving to scrub the air in your car's catalytic converter. Luckily, palladium-107 emits low-energy beta particles. Beta particles are a form of radiation traditionally leveraged for use by researchers to trace the travel of molecules in test subjects, an area where recovered palladium-107 could one day find use.[13]

If, as in the case of ruthenium and rhodium, the retrieved metals are successfully separated but placed into the supply line before the isotopes sufficiently decay, there is a significant chance that other metals coming in contact with the radioactive salvage would become radioactive as well. Enduring reprocessing and several years of storage is a costly alternative to mining, but it is an alternative method nonetheless, and one that may extend the use of existing materials. Should borders close and trade routes cease hundreds of years in the future, recapturing metals in this manner could go a long way to making a country or continent self-sufficient. A metric ton of fuel rod waste contains four to five kilograms of recoverable rare metals, making the effort worthwhile in dire circumstances.[14]

Making use of metals brought into existence through the obsolescence and breakdown of nuclear fuel rods is a silver lining in the often dreary public perception of nuclear energy. When nuclear power accidents occur, the events capture the attention of the twenty-four-hour news cycle and quickly enter the history books, but as a whole, nuclear energy is quite safe. The added value through the acquisition of useful rare metals with the proper processing of nuclear materials at the end of their service is a bonus, a possible positive financial benefit to offset the costs of nuclear waste management.

Recovering is not an easy or clean process by any means, but it is essential because the other option—burying the entire fuel rod deep underground for the next hundred thousand years—is shortsighted and irresponsible on a timescale that looks beyond the immediate future into the second half of the twenty-first century and beyond. In two to three hundred years' time, if not sooner, it may prove difficult to locate and denote the locations of waste repositories that are currently unknown to the most knowledgeable scientific and political minds.

While supplementing current supplies of useful but already mined metals like ruthenium and rhodium is helpful, the true boon of nuclear processing will be the acquisition of metals that do not exist naturally in the earth's crust or that exist in quantities so small that efforts to extract them would remove any financial incentives.

Where nuclear fuel reprocessing will overcome the threefold hurdles of effort, vulnerability, and cost is in opening up the world to a broad array of illusive elements—all of which are metals—that exist only as products of nuclear fission. Sift through all the rock formations on all the continents of the world, and you will spend the rest of your days without finding a mineable ore containing each and every one of the seventeen elements following uranium—the transuranic elements. While the half-lives of these elements make the six-and-a-half-million-year-long half-life of palladium-107 look like an afternoon siesta, any samples of these elements in the earth's crust have been steadily decaying for four and a half billion years. On the timescale of our planet's existence, naturally occurring samples of the transuranic elements have long since released their pent-up energy and decayed into more stable elements and, in the process, erased all traces of their being.

Our current samples of quantities of curium, berkelium, neptunium, californium, mendelevium and more are manufactured through high-energy collisions performed in a laboratory setting. The world's best physicists and chemists use hundred-million-dollar instrumentation and costly reagents to create increasingly large elements, but due to small yields and high costs, most of the elements never see uses developed further than those brimming from basic property observations and academic curiosity. Plutonium is the glaring exception, but it shows what men and women working across the globe can do with nearly unlimited resources and a defined enemy. While lacking the technical precision, these same reactions go on every day within the fuel rods of nuclear energy reactors. If nuclear waste reprocessing is leveraged properly in the coming century, these elements could be isolated in large enough quantities for us to make use of these mysterious metals in our everyday lives.

NATURAL NUCLEAR REACTORS

Nuclear reactors are not constricted to the cement walls and cooling pools of power stations or military-industrial-complex laboratories. A fortuitous congregation of geography and circumstance has led to the formation of a handful of "natural" nuclear reactors across the planet. Energy, of course, is not harnessed from these underground energy producers, but we can reap the benefit of their work by unearthing the transmuted elements left behind over the course of millions of years of nuclear fission reactions.

The best-known natural nuclear reactor is located in Oklo, a region near the western coast of Central Africa known for its rich uranium resources. A portion of these uranium deposits underwent nuclear reactions, but neither the government nor research scientists in this region had anything to do with them. Under the surface of Oklo, uranium and circumstance mixed with proximity and volatility to create a bizarre phenomenon: a natural nuclear reactor. The world would not learn of the natural nuclear reaction until 1972, when the French Atomic Energy Commission (now the French Atomic Energy and Alternative Energies Commission) delved deeper into an analysis of ore taken from Oklo. The ore showed far lower concentrations of uranium-235, the isotope of uranium used to generate energy in nuclear reactors, than normal.[15] Through the decade of the 1970s, the picture of Oklo's past came together, showing Oklo to house *sixteen* natural nuclear reactor sites operating five hundred million years ago, nearly three hundred million years before the rise of the dinosaurs.[16]

What happened at Oklo defies the odds. Nuclear reactors created by some of the brightest minds in the world and financed by the wealthiest governments operate with large trained staffs working in the face of ever-present danger. If not maintained carefully these reactors can and, as we have seen in the remnants of Three Mile Island, Chernobyl, and Fukushima, will fail, tearing a swath of death, destruction, and depopulation in the surrounding area. How could Oklo, without any intervention, maintain nearly sixteen nuclear reactors for years, without turning a large portion of what we now know as Central Africa to dust?

As for the short, snarky answer to that question, Oklo could have blown up, and we just cannot tell from our place in time. While the amount of energy garnered through these nuclear reactors would have been dispersed over a large area, a small group of astrophysicists in the Netherlands posited in 2013 that a natural nuclear reaction gone awry led to the ejection of a massive portion of the planet, which went on to exit the earth's atmosphere and become the celestial object we call the moon. Yes, the moon. You will not find me at the local pub placing a wager in support of the validity of this theory, but it does make for an interesting origin story for the moon and explain why samples brought back from its surface are remarkably similar in composition to rocky portions of the earth's crust.[17]

An interesting but simple explanation for the mediation of the nuclear reactions at Oklo exists, one that blends local geological phenomena with what we know about the region's climate. Fallen rainwater, on its path to reenter the water cycle, likely trickled through the natural nuclear reactor sites at a steady rate. This trickling rainwater would, in theory, cool the nuclear reactors and act as a regulation method. When the reactions within the uranium ore bodies grew too energetic, the heat would boil the water and kill the reaction. Additionally, the reactors at Oklo did not create the phenomenal quantities of energy we associate with nuclear reactors; instead, they steadily churned out one hundred kilowatts of energy, or one-hundredth of 1 percent of the energy output of a commercial nuclear reactor.

During the course of billions of natural nuclear reactions at Oklo, individual atoms of the metals we now deem synthetic, including technetium and plutonium, were created as by-products, but we will never find more than traces of these metals within the reactors at Oklo. Any components of either of these elements along with the remaining transuranics created within natural nuclear reactors have long since decayed into lesser elements, with technetium becoming ruthenium, and plutonium decaying to uranium over several hundred million years.

CHAPTER 7

COUNTERFEITING GOLD

Going to war for land or financial gain is not new, but what about scouring the ocean for treasure in order to pay for war reparations? One such scheme involved Fritz Haber, a Nobel Prize–winning chemist under employ of the German government. Haber was the epitome of a German loyalist, a world-class scientist more than willing to lend his expertise to Germany's World War I efforts. Despite this militaristic streak, Fritz Haber is best remembered for the development of the Haber-Bosch process: a simple method of synthesizing ammonia for use in fertilizer that has spared countless lives across the globe from famine. With the aid of his first wife, Clara, a PhD chemist herself, Haber spearheaded the sinister weaponization and dispersal of chlorine gas and other chemical weapons from positions of leadership in the laboratory and the battlefield. Fritz Haber performed these duties without batting an eye, but the same cannot be said for Clara. In despair due to marital problems and ethical concerns over the use of chemical weapons, Clara committed suicide with one of Fritz's military-issued pistols on the eve of the initial deployment of chemical weapons against Russian soldiers gathered along the Eastern Front.[1]

After World War I, the German government tasked Haber with a seemingly impossible mission: find a way to pay their portion of one hundred and thirty-two billion gold marks owed as wartime reparations to Russia, France, and Great Britain.[2] Haber had the ingenious idea of extracting a commodity of value from a common and inexpensive substance, a scientifically accurate method of extracting gold from seawater. Haber found the process commercially viable at first; his initial estimate showing sixty parts per billion of gold in nearby seawater would yield seven ounces of gold from a ton of seawater.

Delighted with the prospects, the German government sent Haber afloat to scour the oceans of the world and obtain hard evidence for his hypothesis. When Haber initiated the recovery efforts, samples showed a critical flaw in his initial calculations. Theory did not mesh with experimental data, and Haber recovered barely one-fiftieth of an ounce of gold from a ton of seawater. Encumbered by a recovery rate less than one-third of a percent of the initial goal, the German government abandoned the project. Haber went on to develop a cadre of chemical weapons used by Germany in World War II only to be forced to flee Germany under Hitler's rule due to his Jewish ancestry, while the German people would continue to pay their reparations bill until 2010.[3]

Retrieving copious amounts of precious gold from nothing? Haber is not the first genius to tackle a scheme such as this. Alchemists attempted to master the practice of aurification, the act of transforming lead and other "base" materials into gold, for several centuries. Texts support the presence of two schools of alchemical thought: one concerned with the physical transformation of materials and another a self-help style of philosophy purported to wield power capable of changing one's life.

Our understanding of modern chemistry and metals owes a tremendous debt to alchemists steeped in Islamic and Greek culture. In addition to creating early forms of laboratory implements like the crucible, early alchemists developed and codified basic laboratory techniques now well known to general chemistry and organic chemistry students like filtration, crystallization, and distillation. These men and women kept alive the practices that would evolve into chemistry as they strove for an eternally out-of-reach goal.[4]

Our understanding of alchemy is still dim. One the one hand, alchemy appears to be a legitimate undertaking, thanks to texts we have preserved detailing the work of a number of Islamic individuals like Jābir ibn Hayyān and Roman-born Arabic philosopher Morienus, both of whose work was revered in the European world well into the middle of the sixteenth century. European readers would go so far as to bestow the title "Father of Chemistry" upon Jābir ibn Hayyān.[5] Ancient alchemists did not succeed in transforming the base into the precious, but their successors centuries later used a combination of physics, chemistry, and a technological revolution to push the impossible within the realm of reason.

ANSWERING ANCIENT QUESTIONS

Thirty years after winning the Nobel Prize for his hand in the geopolitical shaping–discovery of plutonium and three other transuranic elements (americium, curium, and berkelium), Glenn T. Seaborg turned his eye to an ancient alchemical quandary. Is it possible to turn atoms of a lesser element into gold? In 1980, Seaborg and his colleagues performed a set of history-making experiments at Lawrence Berkley National Laboratory, aiming to eject protons from the nucleus of a bismuth atom. An individual atom can exist with varying number of neutrons before energy or matter must be released, but the addition or subtraction of just a single proton is a dramatic change, turning the atom into a different element. Seaborg's tried and tested team set their goal high, eventually succeeding in removing four protons from the nucleus of bismuth atoms in the course of their experiments.

Bismuth is the relatively inexpensive elemental namesake of the gastrointestinal cure-all Pepto-Bismol and holds eighty-three protons in its nucleus. Seaborg essentially played a game of pool at the atomic level and, in jettisoning four protons, created gold, which sits four spots before bismuth on the periodic table. Seaborg and his research team slammed atoms of copper and neon into a bismuth target at extremely high speeds, and they actually succeeded in transforming a minuscule region of a bismuth target into several hundred atoms of gold.[6]

At least in theory, a skilled scientist armed with infinite time and resources could make any one of the elements on the periodic table within a particle accelerator. Expose a sample of mercury to a flood of neutrons at incomprehensible speeds, and you have created gold, solving the mystery of alchemy in an afternoon. Why are nuclear reactors around the country using the fruits of neutron bombardment to churn water and create electricity when they could be making gold bricks? The answer to that question is a convoluted one, but the logic behind the answer is forever linked to a many-faced resource that drives the actions of the world from the tiniest insect to major corporations: energy.

Seaborg's work is not applicable to our everyday lives. Looking out my office window, I do not see any factories crammed wall-to-wall with Seaborg's "atom slammers." Large corporations are not crashing nuclei into one another to create as much gold as we need to coat the innards of every electrical device and to create any trinket a loved one might imagine. If the technology and desire is present, why are scientists and industry failing to turn seemingly useless atoms into ones we deem useful and, in doing so, eliminating the world's problems one by one? Because the amount of energy needed to perform these targeted reactions wherein an element is intentionally transformed into another is mind-boggling. Combined with the fact that these reactions are successful in creating only a few hundred atoms at a time, and the economics of this exchange quickly damn its widespread use.

There are over eighty-six *sextillion* atoms (to give an idea of this magnitude, a single sextillion is a trillion times a billion) of gold in one ounce fit for trading on the international market. It would take the combined efforts of an all-universe army of scientists working around the clock improving already existing Nobel Prize–winning research, but this is not where the work ends. If successful, these experiments would need to be repeated a nearly infinite number of times at minimal energy cost before enough gold to strike a single small coin is created. Even if science succeeded in this endeavor, worries would arise if this industrial atom smashing became "too" efficient—make large quantities of gold at minimal expensive, and financial markets are likely to convulse. As the market fluctuates, the price of already mined gold would drop to the point where the cost of energy to perform the process would be more than the price of purchasing existing gold, with the two cost barriers battling it out until the market determined a new, stable low price. Although world governments are no longer backed by physical gold, their treasuries will forever be tied to its value. If we take this thought experiment to the edge of the financial carnage that might ensue, we could very well see entire geopolitical relationships ripped apart and reshaped in the process. Even though the technology exists to transmute lesser elements into gold and other metals of varying degree of worth, it is infinitely simpler and, in terms of energy, economically far more efficient to scour the planet for natural reservoirs of these metals.

PRETENDING TO BE GOLD

The turn of the millennium and the first decade of the 2000s ushered in, for many, an unheralded degree of social and political strife. Multiyear conflicts rage on both hemispheres of the globe with even the most innocent civilian capable of falling in harm's way. During this era of instability, individuals questioning the basis of day-to-day business transactions backed by a fiat currency system placed significant portions of wealth previously earmarked for bank balances in tangible goods.

In the wake of this financial sea change, gold and silver benefited most from these hedges, with an ounce of gold quintupling in price by the end of the decade, thanks to renewed demand for this historical marker of wealth. The last time the price of gold increased by a factor of five was the twenty-eight years between 1972 and 2000. The rise in precious metals benefited from an air of fear, with many of the individuals hoarding sums of gold and silver also purchasing guns and freeze-dried foods in preparation for a societal collapse that never arrived.

As a growing number of people sought out precious metals, acquiring physical gold and silver reached a level of accessibility not seen since the Middle Ages. "Cash for Gold" and precious metal shops opened in even the smallest towns in North America; half-pound gold bars are now distributed by vending machines in seven countries spanning North America, Europe, and the Middle East; and cable television documents individuals with little professional training stripping the interiors of warehouses in search of gold, silver, and copper to melt down and sell.[7] While value and demand tied to bitcoin, litecoin, and dogecoin and other cryptocurrencies wax and wane with media attention and Internet security scares, historically valued metals including gold, platinum, and silver push on.

Increased acquisition costs often bring about higher prices and new avenues of supply, along with individuals looking to secure ill-begotten gains. Clandestine maneuvers of this sort are performed in the buying and selling of gold with the aid of scientific knowledge about a metal with a handful of similar properties: tungsten.

Iron pyrite, the mineral better known as fool's gold, passes the eye test for gold only to crumble in your hand. Tungsten is iron pyrite's opposite number—the grayish-white metal looks nothing like gold, but when viewed with a scientific eye, it shares several key physical properties with gold. The density of tungsten—the amount of space a single gram of tungsten displaces—is nearly identical to that of gold. Due to similar density values, polished bars of tungsten and gold weighing one pound a piece are so similar in size that it would be impossible for a human to differentiate if not for their different hues. The heat capacity of gold and tungsten are within 5 percent of each other, so human touch would not feel an appreciable difference between the two metals at room temperature. When placed under the scrutiny of real-life applications, the differences between the two metals become more apparent—tungsten remains a solid at temperatures four thousand degrees Fahrenheit higher than the point at which gold melts, while gold is twice as efficient as tungsten when it comes to conducting heat. Gold and tungsten also have very similar prices on the world marketplace, though it takes sixty-plus pounds of the commonly traded ferrotungsten or half a pound of pure tungsten to match the value of a single ounce of gold. While an in-depth analysis is betraying, the face-value characteristics of tungsten make it an inexpensive stand-in for gold.

So, if you are devious and looking for a way to swindle people out of gold, tungsten sounds really great at this point, right? One big problem lies in the path for any would-be gold counterfeiter—tungsten metal is grayish-white, a very different hue than traditional yellow gold. A visual problem such as this can be rectified with willpower and a drill, leading gold-adulterers to hide tungsten metal within solid-gold objects to create a passable fake.

Gold bars come in an endless array of sizes, but the large two-handed bars shown on television and movies and held as reserves in central banks weigh in at an astonishing four-hundred troy ounces and retail for well over a half a million dollars. Counterfeiters are not targeting these large gold bars but rather smaller, ten-ounce and one-kilogram ingots, sizes less likely to undergo in-depth inspection prior to a transaction. Reports of precious metal traders learning they were scammed by keen counterfeits of one-kilogram gold bars with newly drilled holes filled with tungsten prior to the transaction are popping up in China, Australia, and New York City, a sordid trend brought about in recent years by the astronomical run-up in the price for gold.[8] The gold removed from the bar then enters the pocket of the driller, while the bar is passed along to an uninformed buyer at its normal face value.[9] Tales of tungsten bars coated with twenty-four-carat gold also swirl, with purchasers learning of their exceptional misfortune when the top layer peels away like the gold foil covering a chocolate bar.[10]

Although artfully made tungsten counterfeits can pass a visual inspection, both the gold-plated tungsten bar and the drilled gold bar could be detected if one has access to the right equipment. Gold traders make use of a variety of instruments to test their wares, but thorough and intelligent inspections are not always common in the midst of high-profit transactions. Tools commonly used by jewelers and retailers are devices originating from physics and chemistry research labs. These dealers use a high-precision balance to detect the mass of a bar of gold and compare it to the mass inscribed on the bar itself. Another item likely in a high-volume gold dealer's tool belt is a portable x-ray fluorescence detector, which retails for close to twenty thousand dollars. These pricey gadgets are used to measure the quality of gold coatings on a surface layer, making the detectors vital for discerning whether the gold plating is high in concentration (i.e., twenty-four carat) or low.[11]

Of these two devices, only the balance is useful for detecting intentional tungsten contamination in gold bars without drilling a hole into the bar, and only in the case of a keen eye. The twenty-thousand-dollar portable x-ray fluorescence detector is of no use if the interior of the bar is altered, while an extremely sensitive balance, in theory, is capable of teasing out the minuscule difference in density a bar containing tungsten would exhibit. The densities of tungsten and gold vary by only two-tenths of a percent, making it a challenge to detect a counterfeit using a balance, but it is possible.

While a balance is vital to any gold dealer to weigh and organize purchases, one relatively inexpensive piece of equipment missing from the inspection inventory of many gold traders could eliminate any mystery surrounding the interrogation of a suspicious gold bar: ultrasound.

Identical in theory to diagnostic ultrasound instruments used to visualize fetuses during a pregnancy, an ultrasonic detector measures the speed at which sound passes through an object. The speed of sound in metals varies depending on the elements within, but with solid-gold bars, there should be only one metal present. The speed of sound waves passing through gold is well established—it is a physical constant like density or conductivity. When tungsten or any other metal is added to the bar, the speed of sound recorded by the ultrasonic detector will change, straying from the defined value for solid gold. The speed of sound through tungsten is nearly twice the speed of sound through gold (seventeen-thousand-plus feet per second in tungsten versus just shy of eleven thousand feet per second in gold—for comparison, the speed of sound in air is just over eleven hundred feet a second), so an ultrasonic detector will record a much higher velocity in a counterfeit gold-tungsten hybrid bar.[12] More advanced ultrasonic detectors yield visual insight to physical anomalies in the bar, which would make any drilled and filled-in portions immediately apparent.

Gold is not the only precious metal involved in monetary shenanigans. In what sounds like a plan from a film starring a less than trustworthy politician, the idea of commissioning the US Mint to create a platinum coin stamped with a face value of one trillion dollars has gained political capital in financial circles in recent years. Why mint a platinum coin with such an absurd face value? For the sole purpose of the US Treasury Department depositing the coin in the Federal Reserve to eliminate the same amount of debt on the ledger and allow the country to borrow more money. Theoretically, the US Mint could use a couple of ounces of platinum to create a token they deem to be worth a trillion dollars, an act of minting magic that multiplies the value of the raw platinum necessary by several billion percent. Whether or not the Federal Reserve, which derives its power from Congress but does not have to obey the government's every whim, would accept this miracle coin is another matter, but, in theory, such a plan would work.[13] While such a scheme will likely never come to fruition—sending the trillion-dollar coin back to the land of lunchtime debate topics for big-idea economists—the controversy does bring attention to the cost of the metals used to create a coin and their assigned values.

When the cost to mint a coin or print bank notes is different—with an extreme case being the mythical one-trillion-dollar coin—from the value attached to the currency, the difference is known as *seigniorage*. A great everyday example of seigniorage is the US penny. The cost of melting down zinc and a smidgen of copper (pennies have gone from being made entirely of copper up until 1982 to less than 3 percent copper currently), parceling it out into discs, stamping the visage of our sixteenth president on the face, and trucking rolls of the coin from the mint averages two per every penny created. In this case, the seigniorage is a net loss for the Treasury Department, as the department loses a little less than a cent on each newly minted penny, and the net loss continues with the nickel, with eleven cents' worth of materials, wages, and machine upkeep going into creating each one.[14]

Don't feel bad, though, for the career prospects of those in charge at the Treasury Department or the federal workers earning a living at any one of four coin-minting locations across the United States (Denver; Philadelphia; San Francisco; and West Point, New York), as the seigniorage for every other coin is overwhelmingly positive. It costs less than five cents to produce a single dime, eleven cents for each quarter (interestingly, the cost of producing a nickel coin and a quarter dollar are quite similar), and twenty-one cents into the rarely seen dollar coin, which, in my experience, only exists to confound travelers as the desired change denomination of choice for overpriced airport soft-drink machines.

Resourceful individuals and coin junkies looking to make

a profit often manually separate pre-1982 pennies from their change (and in extreme cases, large withdrawals of cases of rolled pennies from banks) to take advantage of the 95 percent copper composition. This high percentage of copper sets the melt value (which does fluctuate over the years) at around twice the face value of a penny, making for 100 percent profit every time a cunning coin collector pulls a pre-1982 penny out of circulation. It is doubtful that anyone would get wealthy by adopting this route due to the time and labor necessary to conduct a large-scale operation, but as you are sorting through your change jar, it is not a bad way to spend an extra minute or two.

Questionable currency practices are about as old as coinage itself, with governments and leaders often willing to abuse their power for their own benefit and to the detriment of precious metal trading. Even the Roman Empire experienced a similar trial to what we currently see with tungsten gold bars, with a crisis in confidence among traders developing in the first centuries CE as the physical mutilation of coins became commonplace. Coins minted during this period were not like ours in that they were rarely uniform circles, allowing devious merchants to chisel away a small amount of gold or silver and retain the ability to use the coin as its designated unit of currency.[15]

Greed and human nature blended together over time, spreading this practice to emperors and leaders in the region. This austere group debased currency in a more efficient manner, clandestinely mixing base metals like copper in with silver and gold to increase the quantity of coins produced. An expert example of this phenomenon is in a denarii Mark Antony minted during combat to pay off the fleet hired and the soldiers he led in the Battle of Actium, a naval battle that pitted the forces of Octavian against the combined might of Cleopatra and Mark Antony in 31 BCE. Mark Antony's side bore the burden of defeat, cementing the reign of Octavian over Rome and the ascension to Augustus and first Caesar of Rome. Antony and Cleopatra fled to Egypt, where they would both commit suicide within the span of a year. Antony saw to it that his soldiers received payment during the course of battle, but through a prescribed form. The soldiers received special coins struck with Antony's name and an image of an oar-laden war galley on the obverse of the coin, but the coins themselves consisted of a silver and copper alloy due to the financial problems befitting Cleopatra and Antony at the time. Rumors of the low silver content in the coins spread throughout the region, leading to the active trade of these coins in a game of numismatic hot potato and ensuring that the coins remained in circulation well into the second century CE.[16]

Modern coins are far more difficult to counterfeit, as the milled notches around the exterior would immediately give away any traces of Roman-style coin shaving. The thin slivers of metal are also difficult to hollow out in the manner seen recently in Manhattan with gold bars. While these two characteristics pose a problem, independent metal manufacturers are overcoming these hurdles. One China-based company specializing in tungsten products is now offering a variety of gold-plated items, including coins, bars, and bricks.[17]

All the gold-plated tungsten items are sold as fakes, but they improve upon techniques used in sordid deals of counterfeit bars. These commercially manufactured and advertised "fake" tungsten-core coins are currently seen as a blight by the coin-collecting and gold-trading community, but someone with an ultrasonic or x-ray fluorescence detector could always use one of these elaborately produced plated coins to test the device in question. If you are a pessimist, the fake coins may turn out to be useful if you lack the financial assets to hoard gold and live your life prepping for an imminent worldwide financial collapse or natural disaster. Gold is desired foremost among precious metals due to historical and traditional sentiment. In a rebooted world where those bargaining for goods lack any sort of detection devices, the look and feel of gold may be all you need.

CHAPTER 8
PALE HORSES

Corporations and nations seek out rare and scarce metals for their value, their ability to improve human life. While humanity teases out new uses and positive attributes from these resources, another side of human nature seeks to manipulate these metals for devious ends.

If you are a detective novel aficionado, you may be familiar with one such devious use of the heavy metal thallium. Agatha Christie's 1961 novel *The Pale Horse* features a cast of characters suffering the ongoing effects of thallium poisoning; anecdotal reports later claimed that the details of Christie's tale gave women across the world the ability to call down death with a readily available household chemical. One of the more famous accounts of thallium poisoning comes in "Horseman, Pass By," an article in the July 17, 1972, issue of *Time* magazine. The article tells the tale of a London nurse who observed the effects of thallium poisoning not from her medical training but from clues she picked up by reading Christie's *The Pale Horse*.[1]

Thallium became so popular as a murder weapon that the chemical earned the name "inheritance powder" in the dawn of the Industrial Revolution due to the metal's dubious link to convenient deaths benefiting wealthy heirs. When used for ill intent, thallium is dosed not as a spoonful of metal shavings but in the form of the crystalline thallium sulfate. By itself, thallium metal will not dissolve readily in water, making it difficult to hide this form of the poison in a drink. On the other hand, thallium sulfate retains the poisonous characteristics of thallium while behaving similarly to table salt, sodium chloride, bestowing upon the substance a crystalline appearance at room temperature while making the chemical far more concealable. This form is still quite potent, as less than a single gram of thallium sulfate is enough to kill an adult.[2]

Availability mingled with potency and concealment combine to make thallium sulfate an excellent murder weapon. Prior to 1972, thallium sulfate sat on the shelves of supermarkets across the United States as the main ingredient in commercial rat killers.

Thallium ends life by forcing the body to shut down as it takes the place of potassium in any number of the body's cellular reactions and physiochemical processes. Once ingested, the poisonous compound thallium sulfate dissolves, separating the thallium atoms and allowing the metal to enter the bloodstream. The body then begins to incorporate thallium into molecular-level events needed to maintain proper working order, and that's where trouble begins. Thallium atoms are remarkably similar in size to potassium atoms, and this is a problem for the human body. Potassium is a vital part of energy-manufacturing mechanisms and a gatekeeper for a number of cellular channels. Due to similarity between the size and charge of thallium and potassium, the body confuses the metals and allows thallium to substitute for potassium. Unfortunately, this substitution is a deadly one, leading to a shutdown of a number of delicate submicroscopic events that brings about death in a handful of weeks. Erosion of fingernails and hair loss are two prominent late-stage flags denoting thallium poisoning, with the first signs of hair loss showing as soon as a week after consumption of the poison.[3] If you are poisoned with thallium and do not die from acute kidney failure or its complications within a few weeks, your way of life will likely be changed forever, thanks to recurring dates with a dialysis machine.

Thallium poisonings are not consigned to the literature of nineteenth-century England, with a handful of scandalous murder cases, all involving female culprits and featuring thallium as the weapon of choice.

The availability of thallium sulfate no doubt aided in making murderers out of unlikely suspects. Caroline Grills, a grandmother in her early sixties living in Sydney, Australia, earned a commuted death sentence and the nickname "Aunt Thally" for the thallium-laced cups of tea she used to kill her stepmother and three in-laws.[4] Thallium sulfate became a popular murder tool in the late 1940s and 1950s Australia, so much so that the government banned the sale of the chemical in rat poisons in 1952, decades earlier than the United States.

Making use of thallium to kill is not constrained to the mid-twentieth century. The perceived suicide of a Boston-area computer scientist in 2011 was later determined to be a thallium poisoning committed by his estranged wife, Tianle Li, a successful chemist who worked for a major pharmaceutical company, where she no doubt enjoyed access to a variety of thallium-containing compounds.[5] This means of murder is also still close to the minds of Iraqi citizens, as a 2008 poisoning incident targeting Iraqi soldiers left nine ill and two children dead after they consumed thallium-laced cake left as gifts.[6] The Iraqis are, unfortunately, quite familiar with the aftereffects of thallium poisoning. Saddam Hussein used thallium as tool of choice to eliminate political dissidents. Members of Hussein's intelligence group, the Mukhābarāt, allegedly used the metal to taint drinks offered to prisoners held at the command of the president of Iraq in the moments following their release.[7] Unsuccessful plots designed to poison Fidel Castro and an imprisoned Nelson Mandela also made use of thallium.[8]

Unlike many of the heavy metal element poisons, the effects of thallium poisoning can be reversed with the use of a common dye. Prussian blue is typically used as a pigment for painting and printing. The large molecular size of the dye and a peculiar chemical property join forces to neutralize ingested thallium. When consumed, thallium sulfate separates into positively charged thallium ions and negatively charged chloride ions. Relative to the size of many of the molecules in the human body, Prussian blue is enormous. Within this massive molecule are eighteen cyanide groups ready and willing to use their respective negative charges to engulf positively charged metal ions floating around in the digestive system. When administered as an antidote to thallium poisoning, Prussian blue immediately sequesters thallium atoms and drags them through the digestion tract, where the pigment-bound thallium is excreted through the bowels in the form of blue feces, hopefully long before any severe damage is done to the patient.[9]

The pharmaceutical company Heyltex produces Prussian blue for medical use, selling encapsulated doses under the brand name Radiogardase. The US government actively stockpiles Radiogardase as a first-line therapy in the event of a dirty-bomb attack, as radioactive thallium is hypothesized to be one of the components of choice for would-be dirty-bomb manufacturers. Heading to your local arts and crafts supply store to purchase industrial Prussian blue pigment for your "nuclear preparedness kit" is, unsurprisingly, frowned upon by the Federal Drug Administration.[10] The antidote is not a one-shot, one-kill treatment, since the patient must continue on a costly three-a-day regimen for thirty days to ensure all physiological thallium is eliminated. While the several-thousand-dollar cost and the accompanying blue stools are annoying, the alternative, a painful death through organ failure, is far worse.

POLONIUM TRADECRAFT

A second scarce metal doubling as a popular modern poison is polonium. While polonium never enjoyed the widespread availability of thallium—one must refine uranium deposits to acquire the metal—polonium is at the center of two high-profile, politically charged deaths that took place in the beginning of the twenty-first century. Polonium metal is highly radioactive and as such comes with a twist. Radioactivity does not choose sides, leaving polonium to pose a serious health threat to not only the target but also to the would-be assassin, as we will see in the case of former KGB agent Alexander Litvinenko.[11]

A highly decorated agent at the age of thirty-eight, Litvinenko lived at the junction of the Cold War and modern Eastern European espionage. Formerly an expert spy with the Russian Federal Security Service (formerly part of the KGB during the Cold War), Litvinenko became incensed after allegedly receiving orders from the Federal Security Service to assassinate high-profile Russian citizens, including a businessman for whom Litvinenko performed security work. A bold man, Litvinenko did not keep his mouth closed; instead he gathered four supporting agents and held a press conference to formally accuse his superiors of wrongdoing.[12]

Shortly after the press conference, Litvinenko fled Russia but continued to be a potent source of information for journalists looking for insight to the inner workings of President Vladimir Putin's political machine. Luckily for Litvinenko, the United Kingdom granted the former agent political asylum, and he eventually became a naturalized citizen while receiving compensation for consulting with British and Spanish intelligence agencies.[13]

If Litvinenko drew the ire of the Russian Federation with his bombastic departure and aid from foreign governments, he found himself in their gun sights through claims made in two books he authored in 2002—*Blowing Up Russia: Terror from Within* and *The Gang from Lubyanka*. In the books, Litvinenko claims that his former employer, the Russian Federal Security Service, orchestrated domestic terror attacks to create an atmosphere of fear in the country leading to a groundswell of support for Vladimir Putin in his rise to power. While action movies show espionage agents killed by a bullet to the back of the head or a car "accident," Litvinenko's demise was out of the ordinary. His death came from a new twist on a centuries-old assassination trope—green tea spiked with a radioactive isotope of polonium.

During the course of his daily interactions on November 1, 2006, Litvinenko met with two former KGB agents over a meal and drinks at London's Pine Bar.[14] Litvinenko vomited throughout the night and the next few days, but for unknown reasons, he waited three days to visit a hospital. Physicians made a general diagnosis of poison but were unsure of the source. Thallium was initially believed to be the culprit until a series of blood tests showed no signs of the metal in Litvinenko's body.[15] Liver failure, along with the degradation of bone marrow, killed his red blood cell supply and crippled his lymphatic and immune systems. Images of Litvinenko in his last days of life show the frail image of a once strapping man.[16]

Three weeks and one day from the night of the poisoning, Litvinenko died of a massive heart attack brought about by the failure of multiple organ systems.

As journalists and governments raced to uncover the perpetrator and cause of Litvinenko's death, a postmortem analysis found the radioactive metal polonium to be present at twenty different locations over London, Moscow, and Hamburg linked to Litvinenko and the two dinner companions he entertained on the night of November 1, 2006. Included in the contamination sites is the bar where Litvinenko dined that night; traces of polonium were found in the teapot used to serve Litvinenko and his companions, the wall by which he sat, and in numerous areas in the restaurant's kitchen. Investigators also uncovered traces of polonium at the home venue of the Arsenal football squad, Emirates Stadium, when it was later revealed that one of two former KGB agents who visited Litvinenko, Andrei Lugovoi, attended the game between CSKA Moscow and Arsenal the day prior to the meeting.[17]

Further investigation discovered traces of radioactive material on the aircraft Lugovoi took from Moscow to Great Britain and his hotel room.[18] The evidence surrounding Lugovoi would prove damning to those looking to rule out a connection to the Russian Federation, especially in light of Lugovoi's hospitalization in December 2006 due to radiation poisoning and Russia's refusal to extradite their former KGB agent to the United Kingdom for further inquiry.[19]

Once scientific analysis showed polonium to be the cause of Alexander Litvinenko's death, forensic explorations were made to detect polonium in the remains of one of the most controversial leaders of the twentieth century. Swiss scientists studying the exhumed body of Palestinian leader Yasser Arafat in November of 2010 found nearly twenty times the baseline amount of polonium in his bones, along with traces of the radioactive element in his clothes and the soil where he was laid to rest.[20]

Arafat died in 2004 from what is described as a stroke by his attending physician after a bout with the flu characterized by vomiting—a symptom that plagued Litvinenko immediately after his poisoning. The discovery of such a large concentration of polonium has changed the way historians and political scientists view Arafat's death, this finding fostering a growing movement to paint it as murder by an unknown culprit. This is not the first intimation of foul play surrounding Arafat's death: his former adviser Bassam Abu Sharif publicly accused Israeli intelligence operatives of poisoning the Palestinian figurehead's medicine and placing thallium in his food and drinking water.[21]

While the role of polonium in the death of Yasser Arafat remains unsettled, the similarities between his death and that of Alexander Litvinenko are intriguing. Arafat lived in a fortified compound just outside of Jerusalem for the last two years of his life due to ongoing hostilities between Palestine and Israel, making it difficult for an outside attacker to get close enough to poison him. Testing Arafat's former compound for polonium would be a worthwhile endeavor as numerous traces of polonium across various locations were useful in reconstructing the means of Alexander Litvinenko's poisoning. However, the short half-life of polonium-210—a meager one hundred and thirty-nine days—combined with the intervening years would likely render any polonium signal in such a large environment undetectable.

HOW DOES POLONIUM KILL?

The isotope of polonium used in the poisoning of Litvinenko is polonium-210, a form that emits alpha particles and the occasional gamma ray.[22] These are the same gamma rays that turned fictional physicist and pop culture icon Bruce Banner into the Incredible Hulk. In reality, gamma rays do not bestow a green hue and anger-mediated strength but rather they are packets of energy capable of damaging the cell's information library, its DNA (deoxyribonucleic acid), leading to negative effects on the human body.

Alpha particles, the second form of radiation emitted by polonium-210, are far less dangerous than gamma rays. Alpha particles are not packets of energy like gamma rays; they are the union of two protons and two neutrons, making them essentially an element of helium with both of its electrons removed. Pound for pound, polonium-210 is one of the strongest alpha emitters among radioactive elements, with each alpha particle emission damaging DNA, chromosomes, and ionizing molecules within cells.

While polonium contamination sent a healthy international spy to death's doorstep within a couple of weeks, no one else died as a result of being at any of the twenty-plus contamination sites. Why? The answer lies in the method by which an individual comes into contact with the radioactive metal. It appears that Litvinenko is the only individual who ingested polonium-210 in the weeks surrounding November 1, 2006. Andrei Lugovoi, the KGB agent alleged by many to be the culprit, appears to have spent a significant amount of time in the presence of radioactive materials, including polonium-210. His recovery from radiation poison in late 2006, however, suggests that his exposure, while extended, was far less than the exposure Litvinenko received through direct ingestion. If we break down the types of radiation emitted by polonium-210, we quickly see that the radiation is relatively easy to protect against, at least as long as an individual does not ingest or inhale the metal. Most of the havoc wreaked by polonium-210 after ingestion originates in the alpha particles emitted by the isotope, as the gamma rays produced by polonium-210 are too infrequent to pose a major threat.

Alpha particles are easy to block, decreasing the exposure threat posed to the thousands of people who were working in and around the contamination sites in the weeks following Litvinenko's poisoning. When the outside of your body is exposed to a radiation source in a medical office, a research accident, or by eating at your favorite restaurant that just happens to be the site of a clandestine assassination attempt, you at least have a fighting chance compared to exposure through ingestion. With the radioactive material confined to outside the body, some of the radiation is blocked by layers of clothing, skin, paper, or plastic. A piece of copy paper is enough to prevent alpha particles from penetrating the outer layer of human skin.[23]

Ingesting radiation, as Litvinenko appears to have done through a cup of radioactive tea over dinner at the Pine Bar, removes any of the typical passive boundaries against radiation damage and leaves the tissue exposed without protection and thus intimately close to the radiation source. This is why the radioactive tea proved to be deadly, while the tea kettle containing the tainted green tea provided plenty of shielding to protect the waiter and any other individuals who came in direct contact with it.

Also favoring the survival of innocent bystanders is the inverse square law, a rule of physics detailing how quickly the strength of radiation energy falls off with distance from the radiation source. Moving just ten feet away from radioactive polonium decreases the energy one receives one hundred fold, helping to explain why so many people could skirt danger at each of the radiated sites and come away unscathed.

THE LIGHTER SIDE OF HEAVY METAL POISONING

As the English Premier League soccer team Arsenal prepared to move from their nearly hundred-year-old stadium in North London to a new sixty-thousand-seat venue only a few blocks away, stadium owners contemplated what to do with the remnants of the team's historic home. As with North American and European sports teams moving to a new stadium, the owners of Arsenal Football Club set out to sell any artifact—even those nailed to the floor—at Highbury that might have sentimental value for the club's die-hard supporters during 2005 and 2006. As construction crews prepared to turn the facade of Highbury into high-end apartments, a plan was in place to sell nearly forty thousand deep-red seats from the art deco–style Highbury stadium to season-ticket holders looking to purchase their chosen seat for the rather reasonable price of twenty pounds sterling (a little more than thirty US dollars), roughly the same price as a low-end ticket to an Arsenal match.

Although the source of cadmium in the seats was never revealed, the metal indubitably originated from cadmium-containing pigments used to paint the stadium seats a deep-red hue to match Arsenal's team colors. Reports at the time suggested the minuscule amount of cadmium contained in the chairs did not pose an acute danger, but the liabilities accompanying prolonged exposure to the toxic metal as the chairs lay in bars and mancaves of fans across London led Arsenal to squash the sale of any seats.[24] While owners of Arsenal Football Club lost out on a million dollars in potential proceeds from selling the game-used seats, the team took the loss in stride. Arsenal is historically one of the wealthiest teams in the English Premier League, with team management spending eighty million pounds on player salary during Arsenal's first season playing in their new home at Emirates Stadium.[25]

At the very least, Arsenal fans can be excited that the traces of cadmium remained in their precious game seats and not in a product that could, in theory, be consumed and enter the human body. In 2010, McDonald's recalled over thirteen million souvenir glasses created to promote the animated movie *Shrek Forever After* due to fears of cadmium contamination. McDonald's distributed an estimated seven million glasses prior to the recall, which stemmed from a fear that cadmium used in paint adorning the glasses could, in time, leach from the surface. The likelihood of the threat was minimal, but the no doubt damning media coverage following just a single poisoning from these children's glasses would dwarf the ten-million-plus-dollar loss incurred by the restaurant chain due to the recall.[26]

Cadmium builds up slowly in the body, depositing itself in the liver and kidneys and causing damage by oxidative stress as it stimulates the production of free radicals and peroxides. As far as heavy metal poisonings go, however, the effects of cadmium are rather mild when compared to the immediate decline in quality of life seen in those who have ingested polonium or thallium. The unfortunate are likely to exhibit flu-like symptoms and intestinal problems, along with kidney and bone damage due to a decreased inability to absorb calcium in chronic cases.[27]

Despite the newsworthy scare, a recall of souvenir glasses would not hurt the fourth installment in the *Shrek* franchise, which went on to gross three-quarters of a billion dollars in worldwide ticket sales.[28]

In yet another foray of cadmium into pop culture, an episode of the television series *House* pinned a panoply of maladies exhibited by a Major League Baseball pitcher on cadmium ingested through marijuana smoking. The episode is in part responsible for a groundswell of belief among partakers of cannabis that they, too, might succumb to the same fate. Although cadmium is found in traditional cigarettes, detecting an elevated presence in marijuana is difficult. While a single study shows the concentration of cadmium (and a handful of other heavy metals including chromium, iron, and arsenic) to be higher in the weed of cannabis than its seed, no data currently exists tying cadmium poisoning to long-term consumption of marijuana.[29]

METALS THAT ARE NOT ACTIVELY TRYING TO KILL YOU

While the James Bond–like tales of poisoned assassins and tainted children's cups capture headlines, obscure metals are not out to kill you. A number of scarce metals are used in groundbreaking medical procedures, including the historically undervalued precious metal platinum.

The title "wonder drug" is thrown around frequently in the pharmaceutical world, but a small-molecule drug that can effectively treat lung, ovarian, bladder, cervical, and testicular cancer with fewer side effects than radiotherapy? The integration of platinum atoms in a small molecule to create a drug yields a tool effective at treating a wide variety of cancers. Cis-diamminedichloroplatinum(II), which moonlights as the much-easier-to-say trade name cisplatin, is a simple molecule at the forefront of cancer treatment starring a single atom of platinum at its core.

Structurally cisplatin is a quite simple molecule featuring chlorine, nitrogen, and hydrogen oriented at ninety-degree angles around a platinum core. Making cisplatin is not difficult; the reaction requires only four steps, with the difficulty of the synthesis on par with a typical lab session from an undergraduate student's sophomore year.[30] The high cost of the platinum materials, however, keep the metal out of the teaching labs of even the most wealthy universities due to perceived waste and the thought that a devious lab student might run off with a bottle of platinum tetrachloride in the hope of purifying the platinum metal within.

The discovery of cisplatin's important role in the war on cancer came about as many great scientific achievements do—by complete accident. In a 1965 study of *Escherichia coli* bacteria—the fecal matter component and model bacteria most often used by researchers—a trio of Michigan State University scientists observing the impact of electrical fields on bacteria noted that their cell samples quit replicating, an outcome that failed to correlate with their experimental logic. Like all good scientists, the researchers went into detective mode and began mentally dissecting every part of their experimental setup. Their in-depth look revealed that the platinum metal used in the electrodes to create their experimental electrical fields was being leached slowly into the bacteria's growth medium, inadvertently dosing the bacteria with platinum and causing the *E. coli* to grow to phenomenal sizes and bypass the life checkpoints that would trigger a fission process to create new cells.[31]

While the trio did not come across any interesting happenings when they placed their precious *E. coli* in a variety of electrical fields, they did discover that platinum could prevent bacteria from reproducing. The finding was warmly received by the medical world and led to the incorporation of cisplatin in cancer treatment by the end of the next decade.

Cisplatin brings about apoptosis in cancer cells shortly after reacting with the cell's DNA. Once bound to DNA, the information-carrying molecule becomes cross-linked and thus unable to divide—a step necessary for the cell to undergo its form of reproduction: fission. If tumor cells cannot reproduce, the runaway train of unbounded growth is halted. Cisplatin's effect on DNA can also have another cancer-fighting effect—the wholesale destruction of cancer cells. Cells can stimulate the repair of DNA after determining that it can no longer divide, however, once the repair efforts are unsuccessful—thanks to the presence of cisplatin—the cell starts its own self-destruction sequence—apoptosis—resulting in the destruction of the tumor cell. If apoptosis can be successfully triggered in enough cancer cells, the tumor will begin to shrink.

Patients given cisplatin and two other drugs making use of similar platinum chemistry to achieve the same result—carboplatin and oxaliplatin—experience fewer side effects than those who are treated with radioactive materials, making the pharmaceutical a great option since it gained approval from the Federal Drug Administration in 1978. The popularity of platinum in cancer treatment led medical researchers to investigate the possibility of antitumor properties in rhodium and ruthenium, metals often used in conjunction with platinum in catalytic converters, but with little success due to unforeseen toxic effects not observed with cisplatin. Rhodium does have a medical application, however; the metal is used in medical mammography.[32]

Platinum even finds its way into what many adolescent boys consider one of the great technological advancements of the twentieth century—breast implants. Minuscule amounts of pure metallic platinum are used to "firm up" silicone, giving silicone oil the gel-like feel preferred for the implants. It is not feasible to remove all the platinum catalyst needed in the reaction; around half a milligram of platinum (a few pennies' worth of the precious metal) is left in the average individual with breast implants. This is not good news to some physicians and implant recipients since the presence of platinum arouses fear that the metal could leak from the implants and into the recipient's body. While platinum metal is very stable and not likely to react with anything in the human body, a different, plausible form of platinum, a platinum salt, could cause health problems if present in large quantities. The first scientific report of platinum leaching from breast implants is now maligned due to questionable experimental tactics and the quizzical choice of soybean oil to simulate the interior of silicone breast implants, but concerns over safety do persist.[33]

Thulium, one of the seventeen rare earth metals (and one of the rarest, period), is a possible candidate for use in brachytherapy. Brachytherapy is a form of radiation treatment wherein the metal of import, in this case thulium, is placed in proximity to the tumor. The thulium-170 isotope emits x-rays and has a relatively short half-life, allowing for the placement of a steadily firing atomic gun in the vicinity of cancerous cells, particularly in the case of prostate cancer.[34] Use of thulium-doped lasers in surgical procedures to alleviate bladder cancer shows promise over traditional methods, but this research is in its infancy.[35] Modern scientists are even capable of finding a beneficial use for thallium, the key chemical component of "inheritance powder." While the substitution effects of thallium are fatal—dosing a patient with a minute amount of radioactive thallium allows physicians to perform myocardium perfusion imaging tests—studies used for the diagnosis of blocked arteries—and follow the movement of thallium-tinted blood through the body.

The use of scarce metals does not end with lifesaving medical treatments. Everyday dental procedures are now making use of their unusual properties. Palladium, the only metal outside of gold, silver, and platinum deemed a precious metal and traded on foreign currency exchange markets, is a popular material used in restorative dentistry. Palladium is chosen because the metal exhibits similar properties to and costs less than gold, the metal historically used to fashion crowns. A rarely applied physical property, specific gravity, rears its head to give palladium the edge here as well. Palladium's specific gravity is little more than half the value for gold, allowing a palladium crown with the same size and shape as a corresponding gold crown to weigh less, thus filling the same spot in one's tooth while using far less metal.[36]

Erbium, one of the seventeen rare earth elements, also has a dental application, but one quite different than the structural use of palladium. Lasers constructed from erbium-doped crystals, with the erbium allowing for the emission of light at a wavelength that water readily absorbs, can eliminate small cavities, making drill-less dental work possible. As the laser works to remove the cavity, the emissions also temporarily numb local nerves, removing the need for anesthetic in certain cases and making erbium-doped lasers perfect for the future of pediatric dental work.[37]

CHAPTER 9

GOLF CLUBS, iPHONES, AND TRIBAL WARS

Tantalum is a corrosion-proof metal used to increase the efficiency of capacitors—a useful application that has allowed mobile devices to shrink in size or increase in processing power at a rapid pace in the past decade. Tantalum is found alongside the metals tin and tungsten, with the latter commonly used to make top-of-the-line golf clubs. Sadly, tantalum mining funded rebel factions during the Second Congo War (1998–2003), the bloodiest war since World War II, with five million people killed as a result of the fighting.

In a disturbing nod to the current strife surrounding tantalum, the metal's name comes from the disturbing tale of the Greek mythological figure Tantalus. Tantalus's life was awful—he lived in the deepest corner of the underworld, Tartarus, where he cut up and cooked his son Pelops as a sacrifice to the gods. His sins did not end there, however, as Tantalus forced the gods to unwittingly commit cannibalism by dining on Pelops's appendages. To punish Tantalus for this gruesome gesture, the gods condemned him to a state of perpetual longing and temptation by placing him in a crystalline pool of water near a beautiful tree with low-hanging fruit. Whenever Tantalus raised his hands to grasp a piece of fruit to eat, the delicate branches would move to a position just of out of reach; whenever he dipped down for a drink, the water pulled back from his cupped hand. Mythological lore finishes this mental image of eternal temptation by suspending a massive stone above Tantalus. He was condemned to a world of immense desires constantly within reach but of which he was forever unable to partake, leaving him to perpetually starve against a backdrop of plenty.

The situation Tantalus endured is disturbingly similar to the one citizens of the Democratic Republic of the Congo find themselves in today. These individuals are surrounded by riches—they, in fact, are often responsible for the labor that brings those very riches to the surface—but they are unable to reap their rewards.

Tantalum is deposited within coltan ore, and throughout the Congo it is rivaled only by gold in the eyes of miners. Everyone is the region is aware of the demand and knows that a tidy profit can be made by digging up coltan and selling it to upstream purchasers. The locals can use the money earned to fund anything they want, even war.

STUNTED GROWTH

A laundry list of wars have occurred in the region over the past two decades, but to understand the current plight of the people in the Congo and how it impacts the supply of several high-demand metals in the world, we must look to the mid-1890s and an era when the Congo was under Belgian control.

King Leopold II of Belgium treated the region as his play toy, exploiting the people and exporting rubber and other goods to Europe in an act that would foreshadow the pillaging the Congo has endured in each intervening decade. While the commercially desired resource changed—first rubber, then gold, then an alphabet soup of metals—the well-being of those living in the region consistently takes a backseat to the land's riches. With Belgium's entry into Africa, overwhelmingly Christian influences were pressed on the people of the region.

Belgian influence weakened throughout the 1900s, until an exchange of power officially handed control to the Congolese people in 1961 after a series of small rebellions boiled over.

Lagging behind in education as they began their journey into the second half of the most technologically dependent century in the history of humankind, the people of the Congo were left to fend for themselves, with their land often used as a staging ground for passive-aggressive conflicts between the USSR and the United States acted out by propping up political leaders with a desired set of ideologies. This environment fostered corruption in government at the national and local levels, leading to dissension between neighboring countries.

While the states and provinces in North America as well as the countries comprising the United Kingdom and the European Union lean on each other for support and the improvement of their citizens' lives, the small nations in this part of Africa rarely express such sentiment.

SEASONS OF WAR

Four seasons of extended hostility—the First Congo War, the Second Congo War, the Ituri Conflict, and the Kivu Conflict—centered on tensions between nations and ravaged the Congo in the 1990s and the first two decades of the twenty-first century. Opposing forces in all four conflicts routinely pillaged the prized gold and other rare metals of the Congo to fund their efforts, robbing the region of its prized resources in order to destroy neighboring assets.

The Democratic Republic of the Congo was absent from the early days of the First Congo War since it did not come into existence until the end of the conflict due to a quirk in political lines. Zaire, the nation that would become the Democratic Republic of the Congo, lived in the shadow of the Rwandan genocide, a tragedy wherein the Hutu majority killed nearly a million Tutsis in 1994. As one can imagine, hostilities between the two groups escalated as displaced Hutus and Tutsis found themselves living in close quarters within the borders of Zaire. The Alliance of Democratic Forces for the Liberation of Congo, a rebel group opposing the stagnant government of Zaire, took advantage of the regional instability to reshape the political face of the country. After six months of fighting, the destabilized government of Zaire was torn apart and reassembled into the Democratic Republic of the Congo.

The First Congo War was followed just a year later by the Second Congo War, a war that involved nearly every country in Central Africa: Zimbabwe, Namibia, Angola, Chad, and Sudan—neighbors of the Congo to the north and south—aiding the newly formed Congolese government in what became one of the most confusing wars in recent history due to the sheer number of nations and militia groups involved. Uganda and Rwanda—combatants from the First Congo War—supported rebel factions emanating from within the Democratic Republic of the Congo as they fought against the standing government. From a cynic's point of view, it is easy to see the interest of surrounding nations as simply a violent method of securing resources in the Congo, but the Second Congo War also involved efforts to forcefully restructure the government of Rwanda. The confusion did not make the battles any less real, however, and by the time peace treaties were signed and foreign armies began a multiyear withdrawal process, an estimated five million, four hundred thousand people died directly or indirectly as a result of the war.[1]

Fighting in the Second Congo War did not end all at once, leading armed hostilities raging on through what we now call the Ituri and Kivu conflicts, as well as pockets of ethnic fighting. The majority of combat in all four conflicts took place inside the borders of the Democratic Republic of the Congo, besieging the land and people with decades of fear and trauma, while leading nations of the world sat back and watched from afar. The battlefields and bases of these conflicts are often many miles from the nearest traversable road, making it difficult to acquire not only ammunition but also food and potable water. This modified-resource war is taking the lives not only of its participants but of the wildlife as well. Reports of the capture and killing of gorillas found in the Democratic Republic of the Congo swirled during the height of fighting, with the protected animals taken not so much for their body parts that could be harvested and exported for monetary gain but for a far more primal resource—their flesh.[2]

Fighting returned to the Congo in 2012 with the rise of the March 23 Movement (often abbreviated as M23), a group made up of rebel members of the Democratic Republic of the Congo's army. M23 has its origins in the National Congress for the Defense of the People, an armed militia group that was folded into the country's army in 2009.[3] Unhappy with the path of the Congolese government after peace accords created to end the Kivu Conflict, M23 made its displeasure known by capturing Goma, a mineral-rich city along the eastern edge of the Democratic Republic of the Congo in 2012.[4] Amid dissension within the M23 and concessions made by the Congolese government, the M23 ended its rebellion in November of 2013.[5] Although the M23 has stood down for now, tensions in the Congo are capable of flaring at any moment. We will no doubt see a return to fighting in the surrounding region sometime in the near future, due to the number of rebel groups and long-standing disputes between ethnic groups in the region.

COMPROMISED GENERATIONS

The Democratic Republic of the Congo is far from the only part of the world where rare metals are extracted from the ground through works of human tragedy, but it is the most publicized, thanks to ongoing humanitarian and United Nations efforts in Central Africa.

Accra, the capital city of Ghana, is grabbing the spotlight not for metals mined in its borders but for those taken from its trash heaps. The suburb of Agbogbloshie is becoming a global dumping ground for electronic waste, with the refuse left to rust and decompose with environmental containment measures. The residents of Agbogbloshie are turning to the dumping grounds as a source of income, using low-tech measures to rip any metals of value from circuit boards and cathode-ray tubes in hopes of filling their bellies at the end of the day. Yes, this form of scavenging speeds up the release of toxic chemicals in electronic waste as the disposed laptops and televisions of developed nations are torn asunder to retrieve anything of value, but the most dire consequence of scavenging is falling on the shoulders of those who wade through the dumps because they are exposed to any number of toxic fumes and chemicals. Drinking water and soil are contaminated as rainwater and runoff pool lead and mercury from the dumps, with the long-term effects of dump-and-destroy as it is practiced likely to damage the people of Agbogbloshie for generations to come.[6] Africa is not the only part of the world plagued by this complication of modern life, as electronic waste dumping and salvage is becoming increasingly popular in China, Vietnam, the Philippines, and India.

The bad press associated with tragic operations in Africa and Asia is becoming a stumbling block in the path of regional and global corporations specializing in rare earth mining. Australia's largest supplier of rare earths, the Lynas Corporation, made significant headway on the world scene in less than three decades of existence. Lynas owns and operates one of the largest rare metal reserves outside of China, Australia's Mount Weld, and conducts mining operations in Malaysia and operates a processing center in that country. International politics often guide the future operations of the company, with a sale of majority ownership to a state-owned Chinese corporation in 2010 vetoed internally by Australia's Foreign Investment Review Board.[7] The decision was made out of fears that the sale of one of the largest private rare earth mining corporations to an arm of the Chinese government would negatively affect the ability for countries in the region to purchase rare earths. Soon after the buy-out was shot down, Lynas entered into a long-term agreement with a combination of private and government interests in Japan to maintain an open supply between Lynas and that country.[8]

Gaining a foothold in Malaysia has been difficult but not impossible for Lynas despite the company's apparent willingness to play by the rules. The government of Malaysia welcomed the company with open arms and twelve years of tax-free operation, but the people of Malaysia have not been nearly as inviting. Lynas selected Kuantan, a coastal city with a history of fishing and tourist enterprises, as the home of its Advanced Materials Plant, inciting protests from locals who view the mining and processing of rare earth metals as an environmental death sentence.[9] These protests went unheeded, and Lynas opened the plant in 2012. This battle is a cultural one, with environmental issues and fears that traditional vocations will be eliminated.

Compounding Lynas's ongoing problems in Malaysia is a 1999 decree banning the shipment of nuclear waste to the corporation's home country of Australia. This forces any nuclear waste generated by Lynas operations in Malaysia to stay within the country unless a willing suitor can be found, further frustrating the inhabitants of Malaysia.[10]

In the long-term, the city and people will benefit from Lynas's operations in the region, no matter the burden of these growing pains, as the citizens transition from a hand-to-mouth working life in fishing and other traditional industries to the far more stable income stream accompanying ventures in the rare metal industry. While the situation in Malaysia is the approximate baseline for most of the world, it would be a welcome outlier in the Democratic Republic of the Congo, where accountable corporations are few and far between.

LEGISLATING ETHICS

Governments are just now beginning to use their power to prevent corporations from turning a blind eye to exports from Central Africa. While 2010's Dodd–Frank Wall Street Reform and Consumer Protection Act reformed the banking system and perturbed the stock market, the legislation's breadth extended past the walls of finance and into the trafficking of conflict minerals from the Congo. Lacking the legislative power to ban the import of metals from the Republic of Congo, reforms within Dodd–Frank require companies to track the supply chain of conflict minerals and publicly acknowledge the use of questionable minerals obtained not only from the Congo but also the surrounding nations of Cameroon, Gabon, Zaire, and Angola.[11]

By extending the burden of reporting minerals obtained from nations surrounding the Republic of Congo, the Dodd–Frank Act acknowledges one of the biggest problems facing reform in the Democratic Republic of the Congo—these highly desired ores are actively laundered by neighbors in the surrounding nations with the Congolese government wholly unable to prevent it.

Apple Incorporated and a handful of other companies using significant amounts of tantalum, tungsten, and tin from Central Africa did not need federal legislation to mandate ethics in their organizations. A 2010 e-mail from an iPhone user prompted the late Apple cofounder Steve Jobs to deem the use of conflict metals in their electronics a "difficult problem." Jobs candidly admitted that he did not know for sure if the tungsten, tin, gold, and tantalum used in their products originated from conflict-free suppliers despite their best efforts. Sourcing conflict-free materials, a trend that grew out of the fair trade movement, came to the forefront when stories about the friendly e-mail exchange circulated, leading to a groundswell of support for Apple and other prominent consumer electronics manufacturers to elaborately source the rare metals desperately needed for the products.[12]

Apple Incorporated took steps to clear the iPhone from reproach in 2012, publicly announcing the sole use of ethically sourced—a twenty-first-century descriptor for ensuring workers are treated humanely and paid fairly—and independently verified tantalum in their products.[13] Hewlett-Packard and Apple are taking revolutionary steps to verify their sources of tin, tungsten, and gold, which are used across their line of products, as well as publishing lists of smelters within their supply chains on a quarterly basis.

While ethical sourcing benefits the individual, in theory the practice can have deleterious financial effects on a nation or region. As the Democratic Republic of the Congo becomes something of an industrial pariah, the efforts of ethical locals are forever tainted due to the negative practices of a quickly shrinking contingent. In time, their resources will be depleted—it will be one of the great shames of the twenty-first century if the people and infrastructure of the Democratic Republic of the Congo fail to benefit.

CHAPTER 10

THE CONCENTRATION QUESTION

It is not enough for a nation to have a metal present in the earth's crust within its geopolitical borders to enjoy a wealth of minerals—the metals must be present in the correct concentrations to make mining efforts worthwhile. Coal naturally contains uranium—one to four parts per million. This is not a lot of uranium, but it is a quantifiable amount of the radioactive material nonetheless.

A heavy-duty train car like the BNSF Railway Rotary Open Top Hopper can carry a hundred tons of coal, with a hundred similar cars linked together for a total just over ten thousand tons.[1] This run-of-the-mill train sounds a good bit more ominous with a quick calculation using the parts per million of the uranium in coal. After a few minutes of number crunching, the sensationalist could claim that the bituminous coal train is carrying between twenty and eighty pounds of uranium, and this hypothetical individual, in the midst of making a hysteria-inducing statement, would be correct. Although the movement of eighty pounds of uranium across the heartland of the United States resembles a plot point from a spy movie, black helicopters filled with FBI and Homeland Security agents will not be descending on the trains of North America anytime soon, because the uranium is safely split between millions of pieces of coal spread throughout the train.[2]

This is the same dispersal pattern we see with the distribution of rare earth metals in rocks and quarries. During World War II the United States and Germany did not destroy their coal mines to get a small allowance of uranium to use in the building of nuclear bombs—the coal by itself is far more valuable. Instead, these countries looked to well-known deposits featuring high concentrations of uranium to build their stockpiles. Concentrated deposits of metals—often the only deposits worth mining—are created over millions of years. They are born out of situations where physical and geological events occur in just the right topographic locale allowing for accumulation, with a number of different methods of deposition for each metal. For example, gold is found, albeit in minute concentrations, in the world's oceans, which are fed by rivers that wind through areas with varying climates. Metallic gold is a little less than twenty times denser than water, with changes in the temperature of water decreasing or increasing the density of gold (and the water itself) as the river goes from calm to raucous in the midst of rocky rapids. The tiny particles of gold, due to their high density, will fall out of the river water over time, but we see this happen most often at bends and turns in the river, places where the flow is quickly altered. Concentrated gold deposits are often found at these points of change, where the gold gently falls to the bottom of the river and accumulates over tens of thousands of years.

Even if gold is not present in its prized metallic form, forces in the environment are at work to alter the metal. Gold tetrachloride, a salt featuring gold bonded to four atoms of chlorine, is quite toxic to humans and animals, but some species of bacteria are capable of ingesting the tetrachloride form and extruding metallic gold. The bacteria *Cupriavidus metallidurans* thrives in areas with moderate temperatures and high heavy metal concentrations, and when the bacteria comes in contact with an atom of gold that is electron deficient (as is the gold atom in gold tetrachloride), the genome of *Cupriavidus metallidurans* initiates processes that separate the atom of gold from the chlorines in the molecule. These grains of metallic gold are left to "float" around the cytoplasm of the bacteria. As the grains leave the bacteria each one acts as a "seeding" platform where other grains can attach and eventually form a nugget of extremely pure metallic gold over time.[3]

The environment necessary for deposits of iron, lead, and copper must be a little more complex. Modern iron deposits are theorized to have their origins in molecules of iron oxide dissolved along the ocean floor during the Proterozoic Eon, a period from two and a half billion to five hundred million years ago—well before the dinosaurs or even trilobites roamed the earth. During this two-billion-year time span, Earth's atmosphere contained very little oxygen, around 1 to 2 percent compared to the over 20 percent oxygen content we enjoy today. We do know what events occurred during the Proterozoic that spurred the accumulation of oxygen, with one theory suggesting that photosynthetic cyanobacteria in the oceans began oxygenating their surroundings, reacting with iron dissolved in the ocean and then sinking to the seafloor. Such an event would have been repeated for millions of years, with iron-rich bands of sedimentary rock forming as a byproduct—bands we continue to mine today.[4]

The majority of the rare earth metals, including two of the most useful, niobium and tantalum, are found in igneous rock, leading to several theories that place the origin of rocks containing these metals in the slow release of rare earth element–rich magma from chambers deep below the surface of the earth.[5] The formation could have taken place underground as small portions of magma exited the chamber and cooled slowly, or as the magma pushed through the surface and became the lava flows often associated with volcanic activity.

TIME AND CHANCE AT BAYAN OBO

China's available supply of rare metals rivals the material wealth of oil underneath the sands of Saudi Arabia and the Middle East. A crippling share of the planet's supply of rare earth metals is in China—the United States Geological Survey estimates more than 96 percent of the available supply of these metals is centered within its boundaries, leaving the rest of the world to fend for crumbs under their borders or to rely on Chinese-manufactured products.[6]

Fickle government relations with China often muddy the waters, especially when corporations or governments are looking to import its raw materials. Bayan Obo, the source of China's rare earth supply, is a once-in-a-planetary-lifetime formation. Mongolia's Bayan Obo Mining District first became operational in 1927, starting out not as a rare earths piggy bank but as a valuable source of iron.[7] Using Bayan Obo as an iron reserve was not a mistake on the part of Chinese officials. The value of most rare earths was still largely unknown at that point in history, while iron fit the current world and national needs to build infrastructures, ships, planes, and bullets.

The Bayan Obo Mining District is tucked away along the northern edge of China in an artificial subdivision of the nation known as Inner Mongolia. Inner Mongolia is aptly named since it is a strip of China bordering both Mongolia and a short stretch of Southern Russia, a strip dotted in part by the Great Wall of China containing the same hills and mountains ruled by a man born Temujin and later given the title Genghis Khan by his men. The political structure of Inner Mongolia is complex—the area is an "autonomous zone," one of five designated by the Chinese government in the middle of the twentieth century in part due to the events of the Chinese Civil War (1927–1950). Each autonomous zone covers an enormous swath of land—Inner Mongolia is the third largest of these zones in size yet is still twice the size of Texas—with each region allowed to govern its day-to-day affairs, unlike the twenty-three provinces of China. All five zones were initially populated by a single designated ethnic minority—Inner Mongolia, for example, was home to its namesake, while the Ningxia autonomous zone was home to the Hui people, a fascinating Chinese-speaking group with a shared Muslim heritage. The passing decades since the institution of the zones have seen an influx of the majority ethnic group in the country, the Han Chinese, which has changed the original distributions forever in every autonomous zone except for Tibet.

The minerals containing tantalum, niobium, and other rare metals likely accumulated over the course of a four-hundred-million-year span in the Middle Proterozoic period, a geological eon that does not immediately spring to mind when one envisions life on Earth before humans.[8]

Out of the growing pains of this alien Earth came Bayan Obo. During this time, well before the reign of the dinosaurs, our planet was unrecognizable. Earth bore a closer resemblance to its sister Venus than to its modern-day version: a collection of hot swamp-like areas rife with geologic activity enveloped by an unbreathable atmosphere containing an infinitesimal amount of oxygen. Life-forms on this planet were quite small, with the algae and bacteria that ruled the globe soon knocked off their pedestal by larger but still quite simple trilobites as this unprecedented period of accumulated life-forms ceased.

While we will never truly know how such substantial quantities of varied metals gathered in this section of Inner Mongolia, a number of theories are bandied about by geologists.[9]

The shuffling of Earth's tectonic plates and the movement of lava during the periods of geologic tumult that characterized formation of our planet's landmasses is central to the most prominent theories, with the possibility that the movement of magma could have triggered hydrothermal vents that pelted the earth at Bayan Obo with metals brought from deep below the surface.[10] The rare metals present at Bayan Obo, and throughout the world, are found in the repeating, organized forms of familiar chemical compounds. These molecules typically consist of two atoms of the metal joined by three atoms of oxygen, with variations of the number of metal and oxygen atoms present. This odd couple forms a very stable type of chemical compound, the oxide. Thanks to this combination of metal and oxygen, the molecules are readily taken into mineral deposits. This stroke of luck is not without its own problems, however: the metals must be separated from oxygen before we can use them. But this is a small inconvenience to reap a great many resources, with the profits making all the effort worthwhile in the end.

The local government of nearby Baotou oversees the day-to-day events in the Bayan Obo Mining District. A governing relationship like this may appear bizarre to Westerners, but management of Bayan Obo is not unlike that of a US national park, save for one key aspect—the land is being utilized for its resources instead of being sequestered for the sake of preservation. The presence of abundant natural resources in the vicinity of Baotou transformed the city—much like the steel mills, train stations, and cities that sprouted near mining facilities in the United States in the nineteenth century—turning Baotou into one of China's science and engineering hubs over the past three decades. The Chinese government selected a fifteen-square-kilometer parcel in the city to start the Baotou National Rare Earth Hi-Tech Industrial Development Zone in 1992 to take advantage of the uptick in international desire for metals. Significant tax benefits are also given to corporations operating within the zone's boundaries. Individual incentives are dangled to bring talented scientists and engineers into the region, an area less populated and admittedly less attractive to prospective employees than the world-class cities of Beijing, Guangzhou, and Shanghai. One of the more peculiar attractions for prospective employee is the Saihantala Nature Park, an undertaking that is one-part zoo, one-part amusement park, and lies near the border of the zone.[11]

The domestic beneficiaries of Bayan Obo are in a position of power, with tax concessions, government influence, cheap land, and a steady stream of local workers looking for a stable income meeting at a crossroads to secure China's future. On the other side of the equation are the indigenous people of Bayan Obo, many of whom paid dearly during the growing pains of this present-day boomtown. Mining and refining activities supplanted the long history of farming in the region and forced the families working the land to leave Bayan Obo or abandon farming altogether soon after iron mining began in earnest. Dissension among the population is one of the reasons that gaining access to mining operations in Inner Mongolia is difficult, and particularly so for foreigners. In a curious 2009 incident reflecting this attitude, a cadre of reporters visiting mines at Bayan Obo were not allowed to film but instead were given a DVD by local officials.[12]

Bayan Obo is not without its environmental problems—a four-square-mile waste pond sits just outside the city limits of Batou, a reminder of the processing work performed to extract the natural riches found there. While a pond such as this is little more than blight, monetary value can be placed on even the waste created as companies mine and process in the region. Reactions carried out by enterprising individuals could sift through the waste to find less-valuable metals overlooked by large companies, but the practice is currently banned due to environmental concerns.[13]

Despite its vast mineral wealth, Bayan Obo is far from the only reason China rose to dominate the rare earth markets during the first decade of the twenty-first century. Selling at astonishingly low prices is the clever move that made China the undisputed source for rare earths. By taking advantage of the abundant supply at Bayan Obo, Chinese production of these metals all but ran the previous corporate leaders in the United States and Australia from the world market.[14]

Within a decade and a half this economic plan guided countries and corporations to the cheap and available supply of Bayan Obo, soon putting each at the mercy of China's economic and political policies. A brilliant yet simple tactic effectively yielding a sea change normally only brought about through the devastation of a war, but in this case it occurred without a single shot being fired. This brand of economic policy is convincing foreign corporations in Japan and the United States to open manufacturing plants and offices within China's borders in hope of securing favor and a continuous supply of the rare metals they can rely on in manufacturing.[15] Corporations willing to make the jump into China's metal market are also positioning themselves wisely in the event that China radically increases export taxes on its metal supply, an ever-looming possibility that could destabilize market sectors overnight.[16]

Will we see a day when the dependence on China for rare earth metals ceases? Not likely. The supply of rare earth metals could last several decades if not longer if China exercises wisdom in domestic and foreign economic policy. The rest of the world has little recourse in the face of price increases, as any cache of commercially viable rare metals would likely cost more to retrieve than those sold by corporations inside China. Even if countries drew the political ire of China or simply decided to forge their own path by exploring and making use of a newly found untapped deposit of metals within their borders, it could take well over a decade and phenomenal expense before a semblance of self-sufficiency is actually achieved.

THE MAN WHO FORGED A NEW CHINA

Owning substantial assets is one thing, while making proper use of them, as China has expertly shown, is another. In a country with a population of well over a billion people, the responsibility for shaping China's use of rare metal assets stands on the shoulders of a single man. Deng Xiaoping is unknown to most in the Western world, but he was the force responsible for China's thriving rare metal near-monopoly and the air of economic modernization that has taken hold of the country during the past three decades.

Born in relative obscurity, Xiaoping left home for a work-study period in France in 1919. The fifteen-year-old boy excelled in what most of us would call skilled labor, rising through the ranks of the steel and tractor manufacturing plants he was assigned to with an unparalleled knack for installing new machines and repairing older ones. Deng Xiaoping's overwhelmingly positive experience during the time he spent in the work-study program formed the foundation of his national push to encourage Chinese students willing to educate themselves in North America and Europe, a practice now customary for many Chinese students in science and engineering fields. Upon his return to China, Xiaoping joined the Communist militant opposition to the country's Nationalist leader, Chiang Kai-Shek, ousting the Nationalists and playing a part in the installation of Mao Zedong, history's larger-than-life Chairman Mao, and the Communist faction as the ruling party in 1949.

Working behind the scenes, Xiaoping often made himself the target of Chairman Mao's ire and was labeled a "capitalist roader" due his willingness to stand firm in support of Western economic policies. Because of his stance, Xiaoping and his family became targets of a purge during the Cultural Revolution of 1966. His eldest son, Deng Pufang, was beaten by members of Mao's Red Guard and forced to jump from a window on the campus of Beijing University, resulting in paralysis from the waist down. Government agents saw to it that Deng Xiaoping would be unable to visit his son or aid in his recovery, with the elder Deng forced to return to a regular "day job," working in a tractor factory in a similar capacity to that of the French work-study program he'd been involved in during his late teens.[17]

After four years Xiaoping was allowed to return home, where he could care for his son and visit with his family. Deng Pufang's mental acuity and communication skills would help him overcome his paralysis, and he soon become a key advocate for the handicapped, mentally ill, and disabled later in life. Xiaoping did not return to his former stature immediately. A power vacuum created through the death of an army leader with questionable allegiance to Mao opened the door for Xiaoping's return and ascension within the party to first vice premier, a position akin in power and scope to a member in the cabinet of the US president, with the added responsibility of assuming Mao's role should he be incapacitated. He would not stay in this position long—Mao was consistently inconsistent in his opinion of Xiaoping, allowing his wife, Jiang Qing, a political figurehead in her own right and leader of the "Gang of Four," to gather support for the removal of Xiaoping from the day-to-day affairs of the Communist Party.

Finding oneself removed from a political role not once but twice would be very unsettling in any society, let alone a society where families are split and people maimed at the whim of key political figures. Xiaoping would remain in limbo until months after the death of Chairman Mao at the age of eighty-two from his third heart attack in the span of four months.[18]

Jiang Qing became an enemy of the chairman in his final days. Along with her Gang of Four, she attempted to seize power shortly after Mao died in an act that polarized the political scene and paved the way for Xiaoping's return. Less than a month after Mao's death, the all-but-exiled Xiaoping stood as the preeminent political figure in China and re-assumed his position as First Vice-Premier until his ascent to Mao's cherished title of Paramount Leader in 1978.

Waves of change immediately followed Xiaoping's assumption of power. One of his greatest gifts would be a basic one—a modicum of free speech. He gave the people the ability to more freely criticize the Chinese government in an era of reform now known as the "Beijing Spring." Xiaoping soon toured the United States, a diplomatic door opened by President Richard Nixon's 1972 visit to the Democratic Republic of China. With the door open, several major United States corporations, including Coca-Cola, began to place their products on Chinese shelves.

At the center of Xiaoping's movement was a cry to use China's most significant resource—one billion people coordinated under a single government—to perform feats no other country on the globe could imagine attempting due to their own lack of sheer numbers. Manufacturers like Nike, Motorola, and Coca-Cola opened operations in mainland China during the early years of Xiaoping's leadership, with thousands more to join them in hopes of taking advantage of what looked to be a nearly infinite labor base when compared to the rest of the world.[19]

This thawing of international relations paved the way for Xiaoping to meet with British Prime Minister Margaret Thatcher on his home soil. The occasion marked the beginning of talks that led to the 1984 signing of the Sino-British Joint Declaration and the 1997 return of Hong Kong to Chinese authority. Portugal followed suit, with negotiations to return the colony of Macau to China taking place in 1999.[20]

Despite his work to revolutionize the financial underpinnings of Chinese culture and its relationship with the world, Deng Xiaoping's ideology is often summed up, for better or worse, in two phrases uttered in the twilight of his political career. Xiaoping made a stop during his now-fabled "Southern Tour," a series of inspection visits to four cities in January of 1992, just months after the fall of the Soviet Union. Each city was handpicked as a shining example of the success of his economic policy of market reform in changing the way people worked and lived. During a stop in Jiangxi, the same province in which Xiaoping found himself forced into manual labor and separated from his family during the late 1960s and early 1970s, the political figure surveyed the beginnings of a vibrant industry and put its importance to the future of China in perspective, saying, "The Middle East has oil, China has rare earth [metals]." This belief put a stamp on the importance of a thriving rare earth metals industry to China and the need to protect this resource, one that is bringing the entire world to China's doorstep.[21] A second, simpler statement, "To get rich is glorious!" commonly attributed to Xiaoping, has achieved the status of dogma for summarizing the sprinkling of capitalist principles into the everyday lives of Chinese citizens during his time leading the country. While the results of historical forensics posits that Xiaoping never made this statement, the phrase sums up the beneficial impact of wealth in China through hard work and the wise use of resources brought about by his tenure.[22]

While Xiaoping's rule was marred by the tragedy of Tiananmen Square in 1989 and political cronyism, his lasting legacy was his ability to succeed in the arena where Chairman Mao experienced his biggest failures—pulling the Chinese people out of poverty and industrializing the country. Mao's Great Leap Forward campaign failed to pull China away from a pod-like series of village-based agrarian micro-economies. Xiaoping's embrace of a free-market system allowed for supply and demand instead of government brain trusts to delegate production and permitted the country to embrace industrial and technological advancements that positioned China as a near superpower in modern thought. Deng Xiaoping's willingness to pull the Chinese people from their villages and agricultural roots allowed the country to embrace not only its vast human resources but also its natural surplus of rare earth metals, assets that molded China into a global force.

WHERE DID NORTH AMERICA'S DOMESTIC SUPPLY OF RARE EARTH METALS GO?

North America has a few rare earth metal mining sites, with the crown jewel being the oft-maligned Mountain Pass site deep in California's Mojave Desert. The Colorado-based Molycorp has owned and operated the mine since the early 1950s, but the company changed hands several times and eventually became a subsidiary of Chevron prior to a 2010 sale.

The Mountain Pass site looks nothing like the series of caves and tunnels often associated with coal or gold mining. Molycorp's prize, a gem tucked in the middle of the California sprawl and seventy-five miles from the nearest city, is more rock quarry than classical mine, with this hole in the face of the earth growing larger, one transit ring at a time as rocks containing mineral ore are transferred from the bottom to the surface and then to processing plants.

In a lot of ways, the early days of Mountain Pass mirrors those of Bayan Obo. It made its owners the leading US rare earth suppliers, with the mine acting as the central reservoir of rare earth metals emanating from North America in the 1970s, '80s, and '90s. Mountain Pass performed well as the United States' key source of rare metals well into the late 1990s, when two factors led to the closure of the site. China's meteoric rise as a rare earth manufacturer came at the expense of Mountain Pass's supply. Chinese corporations flooded the market with inexpensive rare earth metals, softening the international market for rare earths to the extent that it was no longer cost effective to maintain Mountain Pass. Prior to swift changes in the worldwide rare earth supply-and-demand balance, Mountain Pass came under intense public scrutiny in 1997 after a series of environmental incidents. Chief among these problems were seven spills that sent a total of three hundred thousand gallons of radioactive waste emanating from Mountain Pass across the Mojave Desert.[23] Cleanup of these spills cost Chevron 185 million dollars, sending the United States' most fruitful rare earth metal mine into a death spiral.[24]

Mountain Pass closed its mining operations in 2002, with enough raw minerals stored in reserves to keep Chevron busy processing and trading for a handful of years. The mine stayed dormant until the price of rare earths increased in the past decade, when Chevron sold the mine to Molycorp, which spent an estimated five hundred million dollars to resume operations. A risky move, but one with an underlying sense of wisdom if Mountain Pass could return to its former glory.[25]

Stating that keeping a corporation, its workers, and shareholders afloat in the rare earth mining industry is an arduous task would be an enormous understatement. Mining is a difficult if not damned industry, one where profit margins are eternally slim and political events can change the world stage in a handful of days, if not overnight. Before Molycorp and other mining entities can earn a single dollar, the corporations must find and acquire a mineral-rich site, tear the prized rocks from the crust of the earth, and then carry out thirty-plus refining steps to isolate a single rare earth metal. The financial markets of the world continue to fluctuate the entire time, with minor changes bringing about a sea change in the mining world as commodity prices fluctuate wildly.

For example, what if the state-owned corporate entities of China are encouraged by the nation's government to limit exports to North America and Europe? Prices soar the next morning, quickly eating up every kilogram a company has in its reserves. But what about the opposite scenario—a private mining corporation announces the discovery of an unexplored cache of bastnäsite in Scotland? Prices plummet, and corporations across the world are forced to limit mining and processing efforts to ensure a market glut years in the future will not kill the industry.

If mining rare metals is such a costly, risky, and arduous task, where will we turn in the future when our supplies of tantalum or tungsten run out? It is within the realm of possibility that the metals we need for the smartphones and video game systems of the future will be ripped from the smartphone in your pocket or the PlayStation 4 currently occupying the space next to your television. Developing countries and enthusiastic hobbyists are already entrenched in the world of rare metal recycling, and as we will see in the next chapter, it is neither an easy nor a safe task.

CHAPTER 11

DIRTY RECYCLING

The life of a discarded smartphone does not end when its owner gets a shiny new replacement. Many people cast the old model aside, sending the bundle of metal, battery acid, and plastic to a landfill, while a small, conscientious percentage chooses to repurpose or recycle their old pal.

After dropping off a stash of old consumer electronics at a designated recycling venue, the electronics in close-to-working order are often refurbished and put back into use through retail sale outlets or donation. Those electronics that are in not-so-great condition are de-manufactured (disassembled) to obtain useful, quickly interchangeable components like RAM. Another option requiring far fewer computer experts exists, however, namely, the wholesale destruction of the electronics to obtain scrap metal. This process is often carried out in large warehouses where industrial shredders tear desktop towers and hard drives into inch-long strands of metal that will later be sold as wholesale scrap. Metals worth considerably more than the steel and aluminum shells of consumer electronics are present in these electronics. Gold, platinum, tantalum, and several other rare and valuable metals are used in small quantities in smartphones and computers, but the employee skill sets and time necessary to obtain and refine these metals often makes metal-specific recycling efforts cost prohibitive. A burgeoning group of North Americans and Europeans with spare time on their hands, however, are turning to at-home recycling efforts to obtain impressive quantities of gold from unwanted electronics.

So why are jewelry-grade precious metals used in electronics?

It's a simple answer—using the metals makes your electronics faster, more stable, and longer lasting. For example, gold is a spectacular conductor. As an added benefit, the noble metal doesn't corrode, so gold-plated electronics do not experience a drop-off in efficiency over time. Gold is plated on HDMI cables and a plethora of computer parts in a very thin layer—a thickness commonly between three and fifteen micrometers (there are a thousand micrometers in a millimeter, if it has been a while since you've darkened the halls of a chemistry or physics department). This very thin, very light superficial coating—thinner than a flimsy plastic grocery store bag—is enough to enhance the efficiency of signal transfer, making it worthwhile to use gold over cheaper metals with similar behavior, like copper or aluminum.

The innards of computers are not the only everyday product coveted by profit-seeking recyclers. Catalytic converters increase the fuel efficiency of modern automobiles and clean up the environment by using the precious metals platinum, rhodium, and palladium to transform toxic carbon monoxide into carbon dioxide and water. This trio of metals are electron rich, with their atoms creating a surface that allows transition states of reacting molecules—essentially molecules in puberty—to become new, environmentally friendly molecules. At the end of the reaction, the platinum atom itself remains unchanged, allowing it to aid in the reaction an infinite amount of times if the surface is cleaned and properly maintained (if your car's catalytic converter has ever gone bad, you know very well that the converter becomes quite dirty with extended use). It is rather difficult to recover the precious metals from a catalytic converter without professional equipment, although some hobbyists collect the catalytic converters and keep them intact in the hope of selling them to professional recyclers.

The amateur scientists looking to recover gold and platinum from computer parts are not too different from the elderly men and women clad in socks and sandals who wander along beaches combing the sands with a small shovel and metal detector in hand. There is one major difference between these two groups of treasure seekers, however. Those performing at-home recycling and recovery from computer parts know where their treasure lies; it's just a matter of performing a series of chemical reactions to retrieve the desired precious metals. Gold is head and shoulders the most desired of the possible metals to refine through recycling due to its abundance in consumer electronics.

A number of companies sell precious metal recycling and refining kits on the Internet, with prices starting as low as seventy dollars, provided the amateur recycler already owns a supply of protective equipment and personally manages chemical waste disposal. More expensive kits make use of relatively safer electrolysis reactions—similar to the hair-removal method touted in pop-up kiosks at shopping malls. This slightly safer method brings with it a much higher price tag, with retail starter kits beginning in the $600 range before rising to several thousand dollars. This high price is the cost of doing business for someone with time and (literally) tons of discarded computer equipment to refine, but the cost of entry is a losing proposition for an individual with a few computer monitors in the attic.

Clipping the gold-containing connectors off and selling them directly to a wholesaler adds a middleman, but it eliminates exposure to dangerous acids and decreases the chance of your house blowing up. A pound of clipped and cleaned connectors can fetch as much as $75, while obsolete, gold-containing computer processors often sell for $1 apiece in bulk sales.

Let's be a little bold, however, and take a look at two hobbyists, and learn how they would skillfully recover precious metals from a pile of computer parts.

AMATEUR RECYCLING

For amateur recyclers, the quantity of starting material is of utmost importance. The cost of the chemicals, the danger involved, and the time it takes to retrieve gold from discarded computer parts is substantial—one would never want to carry out a refining job on a few discarded office computers. Unless dozens, if not hundreds, of circuit boards and processors are at your disposal, it's really not worth the effort from a financial perspective, at least in First World countries. Without a large supply of material, one would be better off trying to strain the negligible pieces of gold in Goldschläger. The cinnamon-flavored liquor contains tiny bits of gold foil, with lore claiming the gold flakes decreased the amount of time between imbibing and intoxication. Using a flour sifter or an old T-shirt as a sieve, one can recover approximately one hundred milligrams of gold from a single bottle, all while being treated to a wild night of easy refining.[1]

Two methods are available to the hobbyist seeking to recycle and recover gold from discarded circuit boards, computer processors, and other gold-containing scrap. In the first—what I will call the "scorched earth" method—the scrap is dissolved in one of the most concentrated acids known, and the gold is recovered through a series of steps that alter the properties of the acid solution and cause solid gold to form and fall out of the solution. The second method makes use of electrolysis and is considerably more elegant from a scientific standpoint but involves the nasty combination of acid and electricity.

To better illustrate these two methods, let's consider two urban gold prospectors—Ron and Anthony. Ron majored in chemistry in college, so he has a better theoretical understanding and a reasonable amount of money to acquire equipment. Anthony lacks an advanced science background, however, he isn't afraid to get his hands dirty, so danger and waste are not a big problem in his mind. Most of what Anthony learns is gleaned from back issues

Native gold deposited on the surface of a mineral sample unearthed in Mexico. *Image courtesy of the Vargas Gem & Mineral Collection, Jackson School of Geosciences, University of Texas at Austin.*

NASA Earth Observatory image detailing the heart of mining operations at Bayan Obo, circa July 2001. The area depicted is approximately four miles wide. *Image courtesy of NASA and constructed by Jesse Allen and Robert Simmon, using data from the NASA/GSFC/METI/ERSDAC/JAROS, and US/Japan ASTER Science Team.*

Ghanaian men sift through a dumping site brimming with scavenged electronic waste in Agbogbloshie, an area in Ghana besieged with environmental and health hazards stemming from improper handling of electronic waste. *Image courtesy of Marlenenapoli/Wikimedia Commons.*

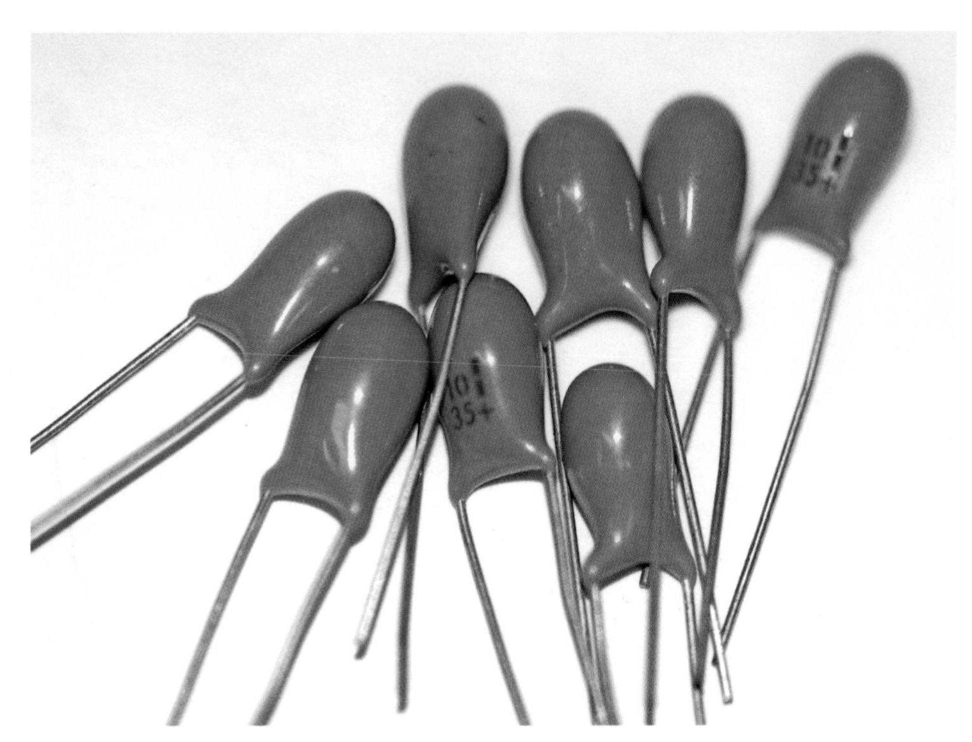

Capacitors constructed using the metal tantalum. *Image courtesy of Mataresephotos/Wikimedia Commons.*

Soviet *Lunokhod*-class rover, the same model of rover purchased by Richard Garriott from Sotheby's in 1993, on display at the Memorial Museum of Astronautics in Moscow, Russia. *Image courtesy of Jason EppInk/Flickr.*

A sample of the mineral xenotime on display at the University of Bonn's Mineralogical Museum (Bonn, Germany). *Image courtesy of Elke Wetzig/Wikimedia Commons.*

A close-up of a manganese nodule, a type of formation that could be key to underwater metal mining. *Image courtesy of Walter Koelle/Wikimedia Commons.*

A close-up of the Lycurgus Cup, a fourth-century CE Roman relic that makes use of rare metals to transform light. *Image courtesy of Carole Raddato/Flickr.*

A small piece of columbite, one of the minerals used to finance conflicts in the Democratic Republic of the Congo. *Image courtesy of the Vargas Gem & Mineral Collection, Jackson School of Geosciences, University of Texas at Austin.*

A sample of the mineral monazite, a source of the rare earth metals neodymium, lanthanum, and praseodymium. *Image courtesy of the Vargas Gem & Mineral Collection, Jackson School of Geosciences, University of Texas at Austin.*

A piece of tantalite, the primary source of the conflict metal tantalum, extracted from East Africa. *Image courtesy of the Vargas Gem & Mineral Collection, Jackson School of Geosciences, University of Texas at Austin.*

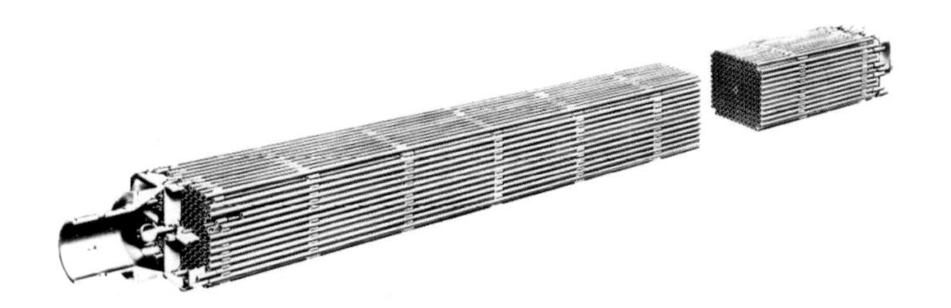

A bundle of forty-one nuclear fuel rods prepared for use in a nuclear reactor. Each alloy-clad rod is filled with small uranium oxide pellets. *Image courtesy of the US Maritime Administration, now part of the US Department of Transportation.*

Workers disassembling electronic waste in the streets of New Delhi, India. *Image courtesy of Matthias Feilhauer/ Wikimedia Commons.*

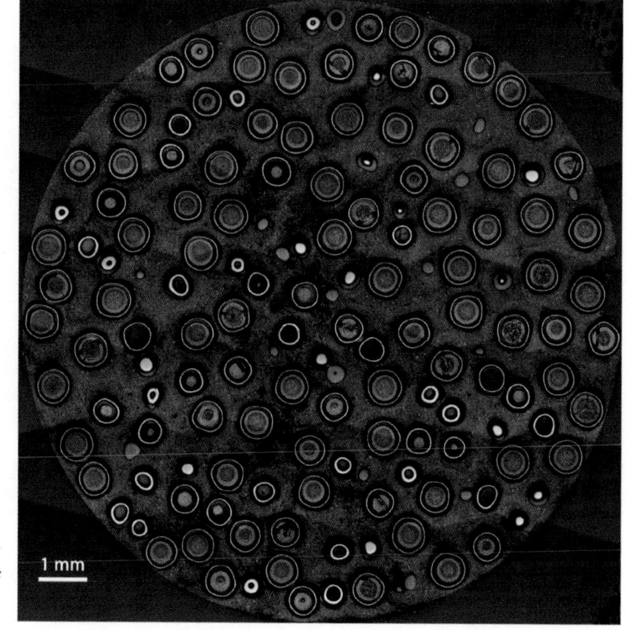

Cross-section of a uranium oxide fuel pellet, the energy-generating source contained in fuel rods. *Image courtesy of the US Department of Energy.*

Personnel cloaked in radiation-mitigating equipment prepare to enter the Waste Isolation Pilot Plant in New Mexico to investigate a possible incident in early 2014. *Image courtesy of the US Department of Energy.*

Miners burrowing through underground salt formations in 2007 to create a long-term storage site for radioactive materials: the US Waste Isolation Pilot Plant. *Image courtesy of the US Department of Energy.*

M. Stanley Livingston and Nobel Prize recipient Ernest Lawrence (circa 1932) stand in front of the cyclotron the two used to observe artificial elements, many of which would be metals. *Image courtesy of the US Department of Energy and the National Archives and Records Administration.*

Miners dig for mineral ores containing tungsten and tin in the Democratic Republic of the Congo, circa 2007. *Image courtesy of Julien Harneis/Wikimedia Commons.*

Aerial view of the mushroom cloud created in Operation Ivy Mike, the hydrogen bomb detonation that created the previously unobserved elements einsteinium and fermium. *Image courtesy of the US Department of Energy and the National Archives and Records Administration.*

A thin sheet of hafnium housed under glass. *Image courtesy of W. Oelen/ Wikimedia Commons.*

Example of the adulterated denarii minted on behalf of Mark Antony to pay those who aided him at the Battle of Actium. *Image courtesy of Professor L. Fontana and Classical Numismatic Group, Inc./Wikimedia Commons.*

of popular science magazines, scouring websites, and personal trial and error.

Let's see how Ron would retrieve gold and other precious metals from computer scrap.

Before Ron begins his extraction process, he crushes and burns the computer parts in a metal drum or trashcan. This is not simply an opportunity for Ron to release some pent-up frustration through the destruction of office equipment; it is also a task that provides him with a reasonably homogenous mixture of gold-containing material. Ron then acquires sodium cyanide and dissolves crystals of the chemical in water. This is dangerous, but not as bad as direct cyanide exposure—such a dalliance will leave you with the taste of almonds in your mouth moments before falling into unconsciousness on the way toward cardiac arrest. Sodium cyanide, in the presence of water, bonds to particles of gold readily, forming a stable complex. Ron will use this property to separate the gold from the computer scrap and pull the gold away from sodium cyanide, leaving him with a very pure gold sample.

Once in solution, an electrical source—possibly a repurposed car battery—is connected to the container holding the sodium cyanide and computer parts. By directly applying an electric current, the gold, in time, will deposit onto the preselected portion of the electrolysis cell (often a steel rod). There are several dangers with Ron's approach—electricity and liquids are not the best of companions—but these dangers are easily overcome by someone with a light background in chemistry or engineering. Depending on the thickness of the gold-plated rod, Ron can scrape or dissolve the metal and start the electrolysis process over again.

Before we move on to Anthony's recycling efforts, let's take a moment and ask a vital question: Why would someone expend the effort, risk damaging the environment, and possibly harm themselves in order to refine metals from scrap? A portion of the draw to hobbyist recyclers could stem from popular fears of impending global disasters and the ensuing financial implications. Gold is sought as a hedge against tenuous economic climates due to its historic significance, but not everyone has the resources to purchase gold coins or gold by the ounce. Many (including Ron and Anthony) do have spare time, and thanks to the Internet a broad range of information is available to individuals looking to recover gold from everyday, obsolete objects.

In Anthony's "scorched earth" method, he dissolves electronics scrap he believes contains gold in *aqua regia* (the name derives from the Latin for "king's water," a historic name given to the mixture because it can dissolve gold and platinum, the so-called "royal" metals). Should anyone partake of *aqua regia*, he or she would surely destroy their esophagus and wreak havoc on their digestive tract, if the unlucky test subject survives.

Aqua regia is a mixture of two strong acids—hydrochloric acid and nitric acid—that can do plenty of damage on their own should they be spilled in the best of conditions, let alone in a makeshift home laboratory. Many of the precious metals, including gold, platinum, iridium, osmium, and tantalum, are highly unreactive metals, and a single strong acid is unable to break them down. Mixing nitric and hydrochloric acid, however, results in a combination by which gold and platinum can be dissolved, lost (but only for a short period of time), and eventually recovered by the hobbyist with a few more steps. It is not an increase in acidity that allows *aqua regia* to dissolve gold and platinum but rather a chemical interaction made possible by the two acids working together. Nitric acid alters gold atoms into a form that will readily bind to free atoms of chlorine, and a phenomenal amount of chlorine becomes available when gold is dissolved in *aqua regia* due to the presence of hydrochloric acid. Gold then exists in *aqua regia* as a stable gold-chloride complex with an indefinite shelf-life. George de Hevesy, a Hungarian chemist working in Denmark during World War II, made use of this interesting phenomenon to prevent Nazis from acquiring the Nobel Prize medals of German physicists Max von Laue and James Franck. Dr. de Hevesy dissolved the two medals in *aqua regia* and left the innocuous container on the shelf of his laboratory for over a decade, until he reversed the reaction and sent the gold to the Royal Swedish Academy of Sciences in Stockholm to recast the awards.

So far, all Anthony has done is toss a cadre of computer parts lined with millimeter-thin sheets of gold into the tenacious acid. Anthony uses better judgment and does not do what some minor-league scientists with a death wish do at this step—heating the acid mixture to upward of one hundred degrees Celsius in order to decrease the dissolving time. Remind yourself, this is all taking place, in the case of a hobbyist, within a garage, apartment, or makeshift outdoor laboratory.

While *aqua regia* eats away at the plastic and metal in the electronics scrap, a number of toxic fumes are released. Anthony had better be wearing a respirator or performing this reaction in a fume hood and clad in acid-resistant coat and gloves. He should be wearing a rugged pair of shoes as well since sneaker bottoms take on the consistency of used gum when you step in a puddle of hydrochloric acid.

Once the mixture is dissolved, Anthony makes the use of the best in over-the-counter protective equipment (hopefully remembering that pair of gloves) to remove the partially dissolved electronics from the vat of nitric and hydrochloric acid. He then washes the partially dissolved parts off in water, collecting any small pieces that fall off and placing them back in the *aqua regia*. So far, Anthony has manufactured two very nasty sources of waste—a vat of *aqua regia* and the remaining water, which is now considerably more acidic than normal and contains a variety of random metals, including the lead and tin from dissolved solder, the "metallic" glue that keeps computer parts together. Never has recycling posed such a threat to the environment.

At the moment, any gold or other recoverable precious metal

is lost forever within highly corrosive acid unless steps are carried out to reverse the process.

Anthony's gold (and possibly platinum, depending on the raw materials used) doesn't look much like gold anymore. In fact, the solution doesn't look like anything worth keeping. No visible pieces of solid gold are present, and the *aqua regia*/scrap mixture has changed from the clear, yellow-red color of *aqua regia* to a dirty-looking opaque green. Anthony shouldn't be discouraged, because his treasure is near: he just needs to perform some chemical magic.

At this point, he adds a chemical to selectively precipitate the metal he is looking to collect. He chooses an extremely inexpensive powder, sodium metabisulfite, since he is primarily looking to retrieve gold. Use of sodium metabisulfite is dangerous; this benign-looking white powder can set off asthma attacks or severe allergic reactions in a percentage of the population. Slowly adding sodium metabisulfite to the *aqua regia* mixture while stirring the liquid is much, much safer than the hypoallergenic alternative. If Anthony has a death wish, he could choose to slowly bubble the extremely toxic gas sulfur dioxide into the solution to reveal solid gold and leave him with an increasingly clear *aqua regia* mixture. Let's nix that idea, on the grounds of safety and insufficient expertise on Anthony's part.

After adding sodium metabisulfite, Anthony waits and hopes to see solid particles of gold—particles now the size of a grain of sand—begin to form in the solution. The gold still doesn't share the appearance of normal gold yet—the metal is presently a brown, mud-like slurry, one that needs to be removed from the *aqua regia* solution. Ideally, removal happens only after the *aqua regia* has been neutralized (either chemically or by digesting enough scrap material), with the brown gold being filtered out.

Once the gold-infused mud mixture is filtered, Anthony can use a solution of stannous chloride to find out if all the precious metals are precipitated out of the solution. Once all the gold is removed from the solution, Anthony then begins to clean the brown, sandy-looking gold. The gold is cleaned by washings with ammonia (more waste) and water (possibly more waste), and the gold still retains its brown, sandy disposition, but we are getting closer. Ideally, washing will remove the remaining impurities, allowing Anthony to obtain a purer final product.

Anthony cannot rest yet, although the gold is stable enough that he can take a break and grab a beer. To account for any excess water absorbed by the gold pieces in the washing process, Anthony kicks up the temperature on a hot plate and carefully heats the brown solid. While the hot plate step serves to dry the brown gold, Anthony will still need to perform a final, insane step to see the yellow shine of this most desired precious metal.

While these methods appear crude, they are not far removed from the steps taken by inquisitive researchers prizing pure samples of rare earth metals in the days before industrial-scale refining. In his journey to acquire a laboratory-grade sample of thulium, a little-known professor at the University of New Hampshire performed one of the great feats of anal-retentive experimental science. In 1911 Charles James reacted the remnants of ore containing thulium with bromine and crystallized the compound in water to obtain a slightly purer form of thulium, repeating the process a reported fifteen thousand times before he obtained a sample of metallic thulium that met his standards for purity. This is an interesting way to spend one's late twenties, but a pastime that provided a firm basis for James's later research, a career in which he was the first person to successfully isolate another rare earth, lutetium.[2]

Now for Anthony's "intentional" fun time with fire and, ideally, a payoff for his effort, risk of life, and the cost of purchasing a variety of deadly chemicals. The now-dry sands of gold are wrapped in a cloth or old T-shirt and soaked in alcohol. An off-the-shelf liquor might work, as long as the proof is high enough, but reagent-grade ethanol from a chemical-supply company would work a lot better. In the final glorious step, Anthony lights the alcohol-soaked gold ablaze with an off-the-shelf blowtorch from a home repair store.

Really.

As the solid melts, the contaminants (including the cloth and alcohol) burn off and exit in a variety of gaseous forms. What remains is the shiny luster of gold—a sample up to 99.5 percent pure—in molten form. At this point, Anthony can let his recovered gold solidify as jagged rocks on his desk, or he can pour it off into a heat-resistant container and create homemade gold ingots.

Anthony's "scorched-earth" method creates a large amount of waste, is labor-intensive, and even when presented in this neutered form (a hearty farewell to the possibility of sulfur dioxide gas exposure), the process is pretty dangerous. Additionally, the protective equipment necessary to safely carry out an ongoing in-house refinery shares many of the signs that warn concerned neighbors of a home methamphetamine lab. Anthony is now a pariah in his neighborhood, but he does have some gold to show for his efforts.

Now that we've looked at two popular methods used by basement recyclers to recover gold, let's ask a vital question—how do the amateurs handle the lingering problem of their efforts, carefully disposing of liters of toxic waste and leftover *aqua regia*? Industrial chemical manufacturers and academic researchers are held to stringent guidelines when it comes to caring for and eventually disposing of harmful chemicals through a "cradle-to-grave" initiative. An at-home operation does not necessarily receive regular inspections, but improper waste disposal (i.e., pouring waste into the sewer or a waterway) can result in jail time and enormous fines.

In the European Union, basement refineries walk the line of legality due to the 2003 Waste Electrical and Electronic Equipment Directive, which classifies almost all electronic waste as hazardous and mandates the separation of no-longer-wanted consumer electronics from the normal household waste stream.[3]

Improperly disposing of toxic waste is a felony in the United States, but you have to be caught first. If you are cited, the Environmental Protection Agency can level fines in the tens of thousands of dollars; however, the fines rarely cover the cost of proper hazardous-waste disposal or of creating an infrastructure to repair damage to the environment. In light of the EPA's nationwide jurisdiction, the enforcement of pollution and recycling often falls to the state governments. The proactive government of California adds a small, upfront surcharge to the purchase of any electronic device with a screen to pay for the device's end-of-life recycling and to raise funds to buffer the environmental impact should the user toss the smartphone or laptop in the trash at the end of its life cycle.[4]

Efforts to offset the environmental impact of short-lifespan consumer electronics are dwarfed by the growing consumption of the same devices in China, Pakistan, and India as these regions modernize. It's a numbers game—the developing regions of the world contain the majority of the world's population, a population every bit as interested in conspicuous consumption as the much-maligned citizens of North America and Europe have been over the past half century.

DIRTY RECYCLING IN THE DEVELOPING WORLD

While the "scorched-earth" hobbyist approaches used by Ron and Anthony are dangerous, the Third World equivalent is disturbingly post-apocalyptic. Venturing into mountains of discarded monitors, desktop towers, and refrigerators, children and teenagers fight over sun-and-rain-exposed electronic parts in search of any metals—even ones the First World discards with every can of soda—for possible resale.

The disproportionate value of a dollar in world trade makes these battles worthwhile, and gives reason to collect copper, aluminum, and scrap steel that First World hobbyists often toss aside in their search for precious metals. A mere two-dollar profit margin on twelve hours' work is not worth the effort in the First World, but this tiny sum is the difference between a tenuous hold on life and starvation for the young workers who flock to these small dumps, ignoring the long-term health risks. Along with any possible precious metals, the young prospectors harvest the copper wire that runs through a computer and crisscrosses circuit boards, the aluminum on the degaussing coils of tube televisions and monitors, as well as copper-containing condensers from refrigerators.

Electronic waste collected for recycling with the best of intentions often goes astray with items earmarked for recycling circuitously finding homes in Third World landfills. In many situations, the collecting organization has little-to-no ability to influence how the donated materials are handled after they are passed on to a third party. From this arises a number of problems since it is typically twice as expensive to refurbish or safely recycle the electronic refuse as it is to transport electronic waste designated for recycling to the shores of another country.

Dumping practices have sprinkled garbage ports of call across the world—Accra in Ghana and the village of Guiyu in China are two prime examples. These ports have also appeared in the Philippines, Vietnam, and India, preying on coastal towns where individuals are in desperate need of income.

Once electronic waste is deposited in the landfills of poor villages, the waste will not stay there for long. Locals in Accra and numerous small towns spread across India and China learned of the possibilities for parts from abandoned computer monitors, televisions, and towers and, like the hobbyists mentioned earlier, took up efforts to retrieve the precious components. In a society where economic prosperity and annual average incomes are measured in the hundreds and not tens of thousands of dollars, the few dollars one might make during a twelve-hour foray through massive piles of rubbish is well worth the effort and risk.

The electronics wastelands littered throughout developing countries could not exist, however, without complicit partners in the destination countries. How do these relationships begin?

BOAT TO NOWHERE

The Basel Convention on the Control of Transboundary Movements of Hazardous Wastes and Their Disposal was created in 1989 and signed by over one hundred and seventy countries in the intervening years, with the goal of preventing the transport of hazardous waste from developed countries to less-developed ones. At the center of the creation of the Basel Convention is a single ship with a hull full of hazardous waste: a cargo ship that illustrates the problems inherent when wealthy nations (and even local municipalities) turn to devious means to discard their waste.

An existing waste-disposal agreement allowed for incinerator waste from Philadelphia to be deposited in New Jersey, but the government of New Jersey began refusing the ash shipments in 1985. As incinerator ash continued to pile up in Philadelphia, the local government contracted the Amalgamated Shipping Corporation and the shipping vessel *Khian Sea* to carry the seven tons of waste ash to a privately owned, human-made island in the Bahamas.

The government of the Bahamas refused the *Khian Sea*'s cargo, and, at the same time, the City of Philadelphia withheld payment of money due for waste disposal. This put those on the *Khian Sea* in a bind; the ship could not return to dock in the United States, leaving the crew to wander the Caribbean in search of a place to dump its ashen cargo.

The *Khian Sea* did eventually find a locale to leave some of its cargo: Haiti. Its crew deceptively described two tons of incinerator ash as fertilizer and successfully dumped a portion of the toxic cargo before the local government discerned the true nature of the deposit. The *Khian Sea* continued to look for a place to leave the remaining five tons of incinerator ash, going so far as to change the name of the ship several times in hopes of finding a taker as the crew journeyed to the Eastern Hemisphere. Unfortunately, the tale of the *Khian Sea* ends in mystery, with over ten thousand pounds of ash suspiciously disappearing during a 1988 journey from Singapore to Sri Lanka. Crew members likely dumped the five tons of ash into the Indian Ocean en route to a port in Sri Lanka.[5]

As the story of the *Khian Sea*'s journey spread, the United Nations motioned to draw up and ratify the Basel Convention. Although the convention is in place, it does *not* prohibit export of waste to less-developed countries. The treaty only requires consent on the part of the receiving country, ensuring that there is a willing caretaker at the port. Due to this loophole, the Basel Convention only acts as a sort of Golden Rule governing hazardous waste transfer. Willing caretakers abound, so the lure of setting up dumping sites at coastal villages continues to exist. All that is needed is for someone in a village, under the guise of a used-electronics entrepreneur, to accept a foreign shipment of waste electronics. He or she might earn a small sum of money for the act of acceptance and possibly recover some working electronics for themselves in the process. The rest can be dumped—landfill space is often inexpensive, if not free, in these parts of the world. As we now know, many individuals are more than happy to go through piles of waste and loot the remaining useful metals and components.

Carrying out business in this manner is essentially a get-rich-quick scheme using a large amount of human capital, a financial undertaking setting up "toxic colonies" throughout the world, destroying not only the lives of the inhabitants but the very land on which the villages rest.

TOOLS OF THE POOR

Those who choose to make a living by retrieving electronic waste from dumps, tearing the equipment down, and refining the rare metals found within them are exposed to many of the same hazards as our hypothetical hobbyists, but on a much higher scale. While inquisitive First World hobbyists like Anthony and Ron refine scrap for fun in their spare time, a recycler in the developing world performs the same work but for twelve to fourteen hours a day and with minimal protective equipment due to the prohibitive cost of respirators, gloves, and goggles. They carry out these activities in an even more dangerous environment as well, exposing themselves to the physical hazards of landfills before the first step of metal recovery begins.

Their tools are often crude. Workers place the metals in clay kilns or stone bowls and heat them over campfires. Heating the refuse loosens the solder present on many electronic parts—solder that is typically made of lead and tin. Children huddle over the fire as the scraps are heated to the point where the solder is liquefied and a desired component can be pulled away for further processing.

The cathode-ray tubes in older computer monitors—an item not even contemplated for recovery by First World hobbyists because of the danger and minimal reward—are boons for profit-seeking recyclers in the developing world. Tube monitors contain large amounts of lead dust—as much as seven pounds of lead in some models—and at the end of these fragile tubes is a coveted coil of copper. While copper is not the most precious of metals, it is valuable due to its many applications, turning the acquisition of one of these intact copper coils into a windfall for a working recycler. Smashing a monitor to retrieve the coil often involves shattering the lead-filled cathode ray-tube, doing a phenomenal amount of environmental damage while covering the worker with millions of lead particulates.

What is done with the unwanted scrap after the useful parts are plucked out is another problem altogether. In many situations, unwanted pieces are gathered into a burn pile and turned to ash, emitting harmful pollutants into the atmosphere. What remains in solid form is often deposited in waterways—Mother Nature's trashcan—and coastal areas. There is rarely a municipal waste system in place to recover the unwanted scraps in these villages, and years of workers dumping broken and burnt leftovers into local streams has contaminated the soil and local water supply. Drinking water is already trucked into the recycling village of Guiyu from a nearby town due to an abundance of careless dumping. Cleaning the water system would likely be too costly and a losing battle if the landfill recyclers are unwilling to change their ways. The physiological impact of recycling electronic waste has been best studied among the inhabitants of China's Guiyu village. Academic studies show children in Guiyu to have elevated levels of lead in their blood, leading to a decrease in IQ along with an increase in urinary tract infections and a sixfold rise in miscarriages.[6]

Many of the young workers flocking to the landfills feel compelled to sift through the electronic waste in order to provide for their elders under China's one-child policy, a policy placing an undo financial burden on the current generation. In addition to complications from lead exposure, hydrocarbons released into the air during the burning of waste have led to an uptick in chronic obstructive pulmonary disease and other respiratory problems, as well as permanent eye damage.

Fixing the long-term electronic waste problem in these villages is a complicated and costly proposition. Apart from a generation of children poisoned and possibly lost, this is a relatively new revenue source, with the oldest of the children involved just now entering their thirties. The area of Guiyu was once known for its rice production, but a decade of pollution stemming from electronic waste dumping and refining has rendered the area unfit for agriculture. The lack of a secondary income source in the area only results in a vicious cycle that funnels more young lives into dirty recycling.

Rejuvenating the soil and water sources after years of contamination would require efforts on par with the science fiction trope of terraforming. Numerous science fiction movies, television series, and novels—including *Dune*, *Star Trek II: The Wrath of Khan*, *Doctor Who*, and *Firefly*—use terraforming as a plot device to make otherwise uninhabitable planets lush and primed for habitation through technological hand waving. Real-life terraforming with current technology is far less ambitious and goes by a much less interesting name: soil remediation. Remediation calls for testing the soil, determining salvageable areas, and excavating the top several inches of contaminated soil and covering the ground with fertile soil brought in from another region. Turning over several inches of soil and replacing it with fresh, fertile soil is a reasonable and cost-effective option. As soil turnover begins to rejuvenate an area, additional steps can be taken to purify groundwater or seed the soil with nitrogen-producing "friendly" bacteria, with the nitrogen produced by the microbes used as a naturally replenishing in-ground fertilizer. Unfortunately, in areas poor enough to fall victim to electronic dumping, it is unlikely a government agency would step in to carry out soil remediation procedures.

AN UPSIDE?

Amid the human tragedies and environmental dangers of amateur recycling, there is an upside to scrap recovery. We may very well run out of usable geological deposits of these metals in the future—or, as we see from the ongoing Congolese conflicts mentioned in chapter 9, the deposits themselves could become too dangerous to mine through corporate means. Recycling rare metals from existing electronics scrap could become a viable option to match the current deluge of consumer electronics in a scarcity-driven world.

For example, tantalum is particularly coveted for its use in electronics. The metal is stable up to three hundred degrees Fahrenheit, a temperature well within the range of most industrial or commercial uses of the element. It works as an amazing capacitor, allowing for the size of hardware to become smaller—an evergreen trend in the world of consumer electronics. Tantalum is also useful for its acoustic properties, with filters made with the metal placed in smartphone handsets to increase audio clarity by reducing the number of extraneous frequencies. The metal can also be used to make armor-piercing projectiles.

A run-of-the-mill smartphone has a little over forty milligrams of tantalum—a piece roughly half the size of a steel BB gun pellet when one accounts for the variation in density between the metals. By itself, this is not a significant amount of tantalum, but a push to start harvesting tantalum and other high-demand, hard-to-find elements from discarded smartphones could create a large, semi-renewable resource of these scarce metals. The task of element-specific refining can succeed, as shown to a limited extent in the examples of Anthony and Ron's search for gold and platinum. Though the venture is cost- and work-intensive, such a process could provide a "last resort" should world reserves become depleted, and it could also offer an alternative to the horrors that have accompanied mining coltan and wolframite in conflict-stricken areas of the world.

Should you ever find yourself with a significant number of discarded computers, plenty of free time, and a willingness to protect yourself and the environment through proper waste handling, the 1940s work titled *Refining Precious Metal Wastes: A Handbook for the Jeweler, Dentist, and Small Refiner* is a classic text that will guide you through your work. Written by C. M. Hoke, an author often incorrectly assumed to be male, the book continues to attract near biblical reverence from amateur refiners. Ms. Calm Morrison Hoke worked as a chemist at New York's Jewelers' Technical Advice Company in the first decades of the twentieth century and became something of an urban legend due to her scientific prowess and a 1915 feature written about her in the *International Socialist Review* focusing on her socialist beliefs and writings. *Refining Precious Metal Wastes* includes well-written, detailed instructions for recovering silver, gold, and platinum from a menagerie of period-specific items, including jacket buttons, used dental fillings and bridges, and silver halide waste created by photographers.

CHAPTER 12

AFGHANISTAN'S PATH TO PROSPERITY

To most of the world, Afghanistan is little more than a battleground—a dry, mountainous land where homegrown soldiers launch shoulder-mounted rockets into the sky against a laundry list of foreign invaders. It is a ubiquitous backdrop for action movies and video games where the mere mention of the country conjures up loaded images of war and strife.

Afghanistan has been a political and military pawn of superpowers for over three decades. The country's geographic position—situated between the Middle East, China, India, and Russia—has placed Afghanistan on the front lines of nearly every military conflict during the twentieth and twenty-first centuries. The United States actively armed insurgents waging war against the Soviet Union during the 1980s, a conflict now seen as a proxy war fought between the United States and the Soviets during the final days of the Cold War. The United States, through the Central Intelligence Agency's Operation Cyclone, spent hundreds of millions of dollars a year and trained over one hundred thousand members of the mujahedeen fighting group from 1979 to 1989 to curtail a Soviet invasion of Afghanistan and further destabilize forces in the region.[1] One of the most effective pieces of weaponry sent by the United States were portable Stinger antiaircraft missiles, an armament that allows a soldier with marginal training to destroy an incoming attack helicopter.[2] If these details sound familiar, two and a half decades later the US intervention and the introduction of the Stinger missile system to Afghani rebels became important plot points in *Charlie Wilson's War*, a

2007 film starring Tom Hanks, Julia Roberts, and the late Philip Seymour Hoffman.

The CIA program of training and weapons transfer was so successful in building a guerilla fighting force that the United States attempted to "buy back" any remaining missiles after the Soviet invasion was thwarted in hopes of preventing the Stinger missiles from being used against the United States in future conflicts.[3] At the turn of the twenty-first century we found ourselves back in Afghanistan, but alliances had swiftly changed, with the United States spending over a decade waging war on Afghani soil in an attempt to root out al-Qaeda's infrastructure and to dismantle the Taliban-controlled government, which came to power in the last years of the twentieth century.

DUBIOUS INFRASTRUCTURE

Despite the CIA's success in rousing a fighting force against Soviet invaders, incursions by the United States in the 1980s did little to improve the day-to-day lives of the Afghan people or prepare the country for the technological revolution of the mid-1990s and on into twenty-first century. Further besetting the people of Afghanistan is the state of their own government, which, for the past three decades has been, for lack of a polite term, a mess.

The Afghani people are troubled by a number of roadblocks to financial prosperity and technological advancement, with many of these barriers put in place by their own government.

President Hamid Karzai banned the use and sale of ammonium nitrate in Afghanistan in early 2010. Ammonium nitrate is a small molecule used as a fertilizer that can also be incorporated into explosive devices. Karzai enacted the ban in the hope of making it more difficult for the Taliban and other groups to fashion homemade explosive devices used to kill NATO troops stationed in the region.[4]

Once denied access to ammonium nitrate, farmers in Afghanistan noted an astonishing drop-off in crop yields, yet they received little to no help from the Afghan government to transition away from the use of ammonium nitrate after the ban. Farmers harvesting a nine-hundred-pound-prune yield the previous year saw their yields plummet to one hundred and fifty pounds after Karzai's ban.[5] A drop in yields of as little as 5 or 10 percent in a developed country would be very damaging to its financial bottom line, but in a country in which 36 percent of its people live at or below the poverty line, the absence of ammonium nitrate is downright devastating.[6] Farmers either had to raise the price of their produce or make the move to illegal opium farming to make a living. The allure of opium is, pardon the pun, intoxicating. Raw opium sells for several hundred dollars per pound, and with a probable harvest of roughly fifty pounds of poppies per acre, the attraction is strong for even the most pious of farmers.[7]

While farmers suffered, the Taliban simply turned to a source not subject to Karzai's ban to construct explosives: potassium chlorate, a chemical used in textile mills across the region.[8] In addition, national and local government efforts to reduce environmental damage continually ran afoul of the Afghan people, including an environmentally conscious ban on the use of brick kilns and an effort to limit automobile traffic in the populous city of Mazar-e Sharif.[9] While their intentions were no doubt noble, the actions were shortsighted and resulted in decreased income for the vast numbers of the less well-off living and working in the city. These are excellent examples of the troubles such a developing country faces as it tries to advance its economy and infrastructure while at the same time doing minimal damage to the environment, a problem that continues to plague Afghanistan as the country tries to make the most of its vast resources.

And when government mandates fail or a situation is in need of an immediate response, there is little money available to develop a solution. Erosion and deforestation are blights on the already parched earth of Afghanistan, turning more and more useful acreage into the desert that already covers the majority of the country. A 2012 initiative through Afghanistan's National Environment Protection Agency set aside six million dollars to fight climate change and erosion, an embarrassingly small sum to dedicate to preserving the farmland that provides the livelihood for 79 percent of the country's people.[10]

A weak electrical system plagues the country as it lurches into the third decade of the twenty-first century. Blackouts limit the access of electricity in a significant portion of the country to a mere one to two hours a day, putting modern necessities like refrigeration out of reach. Industrial efforts are also stymied by breakdowns in the electrical system, with money lost and manufacturing forced to halt production due to frequent electrical outages. If the people and government are unable to pay for basic infrastructure elements, will they forever be relegated to the neglected portion of the developing world?

ILLICIT PROFITS

The pale green and yellow pods of the poppy appear alien compared to the plant's simple red flowers, however, the boldly colored flowers are not the center of millions of illegal transactions a year, nor are they the original source for necessary pharmaceuticals that alleviate postoperative and chronic pain. The sticky substance within the opium pods is the source of these life-altering properties, with the white latex in the pods bled out over the course of days and left to cure. The resulting brown resin is ripe with small molecules categorized as alkaloids: an analgesic able to numb the strongest pain. The gummy brown substance bled from the seed pod can be mixed with calcium carbonate, better known as lime, and boiled to retrieve a brown paste. This paste is morphine, but processing often goes a number of steps further, by producing chemical reactions that slightly alter morphine and create the white powder known as the illicit drug heroin. Opium derivatives originating in Afghanistan travel throughout the world, supplying most of the United States and Europe's heroin supply as the drugs travel the "Silk Road," a system of smuggling operations that takes the heroin out of Afghanistan through a route that weaves through Tajikistan, Uzbekistan, Kazakhstan, and Russia before disseminating it throughout the world by air and sea.

The draw of the poppy is not limited to nineteenth-century opium dens and twentieth-century alleyways where a tattered addict clutches a nearly empty sandwich bag, a metal spoon, and a lighter. Evidence of the plant's elevated status is seen as early as 4000 BCE, with seed pods found with traditional burial items in a Spanish cave system. Assyrians and Sumerians harvested the poppy during the height of their respective empires, making use of the quizzical properties of the white liquid flowing from *Papaver somniferum* to heal the sick. The Romans used it to cater to the needs of the ailing as well, noting that an excessive dose—a planned dosage in many cases—could bring about death.[11]

Pharmaceutical-grade codeine and morphine are no longer sourced for retail sale from the latex of opium seed pods but are instead synthesized from even smaller nitrogen-containing molecules in industrial-sized vats. The switch was made in order to obtain a consistently pure product (i.e., one that meets or exceeds the standards of the US Food and Drug Association and oversight bodies across the world) while limiting the impact of market price fluctuation. While pharmaceutical companies have moved away from the poppy fields for their opiate supply, farmers across Afghanistan continue to grow and harvest the plant for use in a variety of less couth manners.

The opium trade is firmly established in Afghanistan, with its origins, once again, in the Soviet invasion of the country in

1979. Exporting the crop provided a disproportionate sum of money for the amount of land needed and the effort required, a far greater return than growing wheat or raising livestock. This source of illicit income became a necessary evil to support a land in need of money to buy weapons for use against ever-present and well-equipped invaders from the north.

Taliban leader and war hero Mullah Mohammed Omar placed a ban on opium cultivation in 2000 and 2001, claiming that the growth and sale of opium went against tenets of Islam.[12] This ban was successful, as areas directly under control of the Taliban eliminated 99 percent of the poppy trade through cessation and crop destruction. The Taliban lifted its pseudo-theocratic ban on opium production shortly after the United States began military operations in Afghanistan, with farmers instructed to begin exporting the fruits of their harvest as soon as possible. Cultivation and sale of opium rebounded, accounting for more than half of Afghanistan's gross domestic product by the middle of the decade, with the cultivation of six thousand tons in 2013.[13] The Taliban imposed a religious tax on opium farmers—an inflated tithe of up to 20 percent of opium taken to market—and created a phenomenal income base for the organization.[14]

The decision to embrace the harvesting of opium changed the course of the country in a myriad of unforeseen ways. Such an embrace has unusual social effects, including the 5 percent of Afghanis addicted to opium-derived illicit drugs today. The problems associated with illicit drugs in Afghanistan are not contained to opium, as Afghan farmers excel in growing other socially questionable crops, such as marijuana. The United Nations found the country to be the world's largest supplier of cannabis beginning in 2010, with the combination of climate and soil providing farmers in Afghanistan with crop yields three times those of the next-highest producer, Morocco.[15]

At the grassroots level, it is difficult to condemn these farmers for growing and processing opium. A thriving market exists for the product, and, regardless of societal ills, the farmers need to be able to provide for and shelter their families. But what if the fields and people of Afghanistan could be used to harvest a resource infinitely more useful?

INVESTIGATING AFGHANISTAN'S STRATEGIC METAL SUPPLY

Nine years into the United States' war in Afghanistan, the Pentagon released the results of the US Geological Survey operation carried out to observe and catalog the potential rare earth resources in Afghanistan. The fabled 2010 report—already bolstered by rumors of a Pentagon memorandum christening Afghanistan the "Saudi Arabia of Lithium"—revealed a treasure trove of previously unknown mineral resources including gold, iron, and rare earth metals. Early speculation placed a one-trillion-dollar value on the accessible deposits, but there is a substantial problem—Afghanistan lacks sufficient modern mining technology to tackle retrieval efforts.

Separate estimates made by Chinese and Indian interests dwarf the figure, placing the mineral wealth of Afghanistan closer to three trillion dollars.[16]

Along with its rare metal resources, Afghanistan is home to a wealth of sapphires and emeralds, as well as rubies, lapis lazuli, and other semiprecious gemstones. Ownership of the gems often becomes a point of contention en route to resale, with the Afghan government placing numerous restrictions on miners. These restrictions are commonly relaxed or ignored in regions dominated by Taliban influence, with a growing portion of individual miners smuggling the gemstones into Pakistan to avoid taxation.[17]

Knowledge of Afghanistan's mineral riches is not entirely new, at least not to academics, the country's Kabul-based government, and their former Soviet occupiers. Afghan geologists created internal accounts of the country's deposits during the 1960s and '70s, while Soviet mining experts collected reams of data during the 1980s occupation. In the intervening years since their assessment, the information and deposits themselves laid dormant, with mining operations leaving most untouched. Illegal mining operations—ongoing mining efforts that lack government approval—were the only mining initiative profiting from the deposits in Afghanistan, with the border province of Khost becoming a haven for these illegal operations and eventual export into Pakistan.

The wealth reported in 2010 is likely a continuation of the work carried out by the US Geological Survey Mineral Resources Project, which aided members of Afghanistan's sister group, the Afghanistan Geological Survey, from 2004 to 2007, to help the country's government determine a workable baseline of their mineral wealth.[18]

While cynicism often reigns when we look at North American incursion into Afghanistan, this may not have been a solely profit-minded gesture, as the USGS also teamed up with the Afghan government to assess earthquake hazards as well as to catalog oil and gas resources in the country during the same time period.[19]

THE FUTURE OF AFGHANISTAN

Accusations of bribery and a questionable history of awarding contracts have plagued Afghanistan's Ministry of Mines, with Pentagon insiders possibly viewing the country as ill-equipped to handle its own mineral wealth. Lacking a solid organizational foundation and proper surveys of assets, it would be within the realm of possibility for regions of Afghanistan to fall into the cycle of strife and conflict that engulfs a number of mineral-rich South African nations.

Tucked between nations constantly embroiled in conflict and rife with religious and tribal struggles, Afghanistan may have many more days of strife ahead of it.

War in Afghanistan and the Iraqi War were condemned by outsiders as struggles fought to secure natural resources—the act of assessing the quality and potential reserves of such natural resources in itself would suggest an intent to extract or make use of the resources to support an ongoing occupation or an effort to exploit them by other means.

To make the most of the valuable bounty beneath its soil, Afghanistan will require significant aid. Infrastructure aid and a long-term beneficial relationship, regardless of the assisting country, would be of benefit to the people of Afghanistan and the region. For example, a strengthened Afghanistan would provide the United States with an improved geographic buffer to China and Russia while creating yet another staging area for Middle Eastern efforts. The most important strategic role that an open-ended partnership in Afghanistan, regardless of the country, could play is in stemming future escalations of military tensions between a nuclear India and a nuclear Pakistan. The presence of a thriving mining-based economy could switch the tide from the many opium-dominated regions of Afghanistan, thereby curbing the international drug trade while supplying the world with much-needed metals. A modest financial and infrastructure benefit obtained with the help of another nation is better than receiving no benefit at all while having Afghanistan's resources squandered.

CHAPTER 13

LITTLE SILVER

While humanity has prized gold and silver for thousands of years, platinum lacks their esteemed lineage. The precious metal holds applications far beyond expensive jewelry and marketing campaigns for credit cards, allowing one to argue, if so inclined, that platinum is of much more use to today's world than gold.

Our ancestors lent value to gold and silver since the beginning of written history, but unlike these two precious metals, platinum has enjoyed a comparatively short relationship since it was not successfully identified until two hundred and fifty years ago. Akin to debates declaring the Vikings over Christopher Columbus as the first to "discover" the New World, historians argue over the initial discovery of platinum, pitting two individuals against each other.

ONE METAL, TWO DISCOVERERS

During the height of Spanish rule in the mid-eighteenth century, Antonio de Ulloa, a Spanish naval officer, accepted a spot on a French exploration of Peru and Ecuador. De Ulloa received orders to perform the mundane but important task of measuring the distance between longitudinal points at the equator—the meridian arc—to see if his measurement varied from calculations of the distance made in Europe. This task made use of his background in astronomy and natural history to improve knowledge of cartography in South America, but it also gave de Ulloa the opportunity to observe the geography and cultural traditions of the people of Peru, providing the subject matter for one of the two well-received works he penned later in life.

On his return trip to Spain, de Ulloa's ship experienced an untimely meeting with a British vessel. Thanks to the ever-simmering animosities between Spain, England, and France during the period, de Ulloa was diverted to London, where the explorer became a prisoner of the British. Held as their captive, de Ulloa exhibited the traits of a model prisoner—he continued to contribute in his fields of expertise, gaining renown in London in spite of his political status. After a few years of limited captivity, de Ulloa earned the right to return to Spain, but he did not stay in his home country long. His nomadic lifestyle and prior experience gave him an education few could best, placing him in a position to become an emissary of Spain's political presence in South America. During a period of downtime he published a book recording the highlights of his time in Spanish America, and this is how we know of his introduction to platinum. De Ulloa noted a particulate found in riverbeds that drew the ire of gold miners in Colombia and Ecuador. The silver and black specks would not melt in the heat of a kiln. If introduced to molten gold, the particulates would discolor the final purified ingot and lower the value in the eyes of prospective purchasers. De Ulloa noted that miners and metal workers would often throw this metal back into the river, giving the silver and black sand the description it carries today—*platina del Pinto*, a phrase translating to "little silver of the Pinto River," or what we now call platinum.

De Ulloa's observation of platinum would go overlooked at this junction in history, but by no fault of his own. Spain soon sent de Ulloa to Peru and placed him in charge of their cinnabar mines, a series of deadly caverns storing the mercury Spain needed to mine silver deposits in other parts of Peru. Mercury and silver found in Spanish America provided a windfall for the country, easily explaining Spain's motivation to use military and political might to protect their interest in the region.

The true reward for his service came in 1766 when de Ulloa became the governor of West Louisiana, the prize of Spanish holdings in what would become the modern United States. While Louisiana dots our current map as a single, average-size state nestled beside the enormity of Texas, two hundred and fifty years ago the borders of Spanish Louisiana stretched from the edge of Mississippi in the east to Canada in the north and to modern-day Colorado in the west. Spain gained control of this expanse from the French through a secret, unannounced pact that rallied Spanish support against Great Britain and at the same time secured Spain's neutrality in the French and Indian War. The public declaration of the handover of Louisiana took place in 1763, but Antonio de Ulloa would not arrive to govern until three years later. Once in Spanish Louisiana, de Ulloa was slow to act. When he did, the ever-shy de Ulloa did so without success, further infuriating the locals. Economic wounds made during his tenure failed to heal, and a rebellion broke out in 1768 due to de Ulloa's unwillingness to honor promissory notes held by the colonists and issued by their former French rulers. Siding with the French, Native American, and African colonists in Spanish Louisiana would have drawn the ire of his betters in Spain and risked severing his lifeline. While the rebellion forced de Ulloa to flee to Cuba for safety in what would be the end of his governorship and political career, his loyalty to Spain allowed him to take refuge from the world through a posting in the Spanish navy.

If not for his chance observation of platinum, de Ulloa would be just another failed foreign leader installed in the New World by a government with little regard for his personal abilities or the sociopolitical makeup of those he would govern—his installation as governor of Spanish Louisiana is an excellent example of placing a square peg in a round hole.

The hindsight of history allows us to see de Ulloa as a far better historian and scientist than a bureaucrat, an unfortunate example of the bumps along the road that arise in a vaguely meritocratic political machine. We can imagine the anthropological and personal achievements de Ulloa might have garnered had he never been sent away from South America—a place he loved, after all.

De Ulloa proved to have an impressive resume and offered a concrete identification of platinum. But historians have lifted up as his adversary a bewildering sixteenth-century Italian physician and scholar known best for his ego: Julius Caesar Scaliger.

Julius Scaliger entered the University of Bologna at the age of thirty after deciding to walk the path to the priesthood. From what history can discern, Scaliger chose this path not from a desire to serve his fellow humans but from a belief that he could rise through the ranks of the Catholic Church and even become pope. The would-be pontiff claimed ties to Italy's once-powerful La Scala lineage (the Scala name morphs into Scaliger, depending on geography and language), a powerful family that ruled over Verona through a succession of sixteen lords during the fourteenth and fifteenth centuries. Scaliger told wild and likely false tales of his youth to match his asserted ancestry, including a claim to have served as a page for Maximilian I, Holy Roman Emperor, during his teens and twenties, as well as his alleged induction into the Order of the Golden Spur, a papal knighthood, for his military valor.

Questions over Scaliger's actual connection to the La Scala mini-dynasty came under intense scrutiny later in his life—scrutiny bolstered by a scandalous relationship with a thirteen-year-old girl. By the time these rumors surfaced, Scaliger benefited from a cult of personality and body of work that allowed him to overcome the allegations. His scandalous relationship would turn fruitful as well, with Scaliger wedding the young woman when she was sixteen and the pair conceiving over a dozen children, including a son, Joseph, who would revolutionize how historians view the impact of Middle Eastern cultures on humankind.

The oft-maligned Scaliger traveled extensively and practiced science in the steps of long-dead Greek philosophers, choosing to embrace thought experiments in lieu of physical exploration, an unusual choice since most branches of science rapidly focused on empirical study during his lifetime. Once he quit pursuit of a religious post, Scaliger joined a number of military expeditions and began to study medicine.

While detailing an expedition to Central America in his 1557 text *On Subtlety*, Julius Caesar Scaliger wrote about a puzzling metal, one that no fire or familiar device could liquefy.[1]

Taking into account the locations of Scaliger's observation and de Ulloa's notes two centuries later, one can cast Scaliger's reference, albeit brief, as the first one to platinum. Whether or not Scaliger's notation justifies discovery is a point of argument, as the connection to metal working was not apparent from his writing, nor did he lend a name to the metal. Any connection between Scaliger's metal and what we now know as platinum comes through the work of historians and scientists, including Antonio de Ulloa, centuries later.

It is also within reason to proclaim that neither de Ulloa nor Scaliger hold a legitimate claim to the discovery of platinum, as both simply observed its quizzical behavior with little or no record of experimentation on samples of the metal. We would be fooling ourselves to state that either was the first individual to witness these bizarre properties, as we have evidence from de Ulloa's own admission that a number of metal workers contemporaneous to and preceding the duo handled the metal and judged it as tainted gold or silver. General use of the metal we now call platinum dates back thousands of years earlier, thanks to artifacts found in the Egyptian city of Thebes. The foremost relic is the Casket of Thebes, a sarcophagus dating to 700 BCE and featuring platinum alongside gold and silver hieroglyphic inlays.[2] That the positioning of platinum along the inlays suggests the craftsmen at Thebes viewed platinum as a separate substance and not as contaminant, however, has not been proved.

ACCIDENTAL INCLUSION

A perfect example of humans using the properties of a metal with little or no scientific understanding is seen in the Lycurgus Cup, a fourth-century CE relic originating in the Roman Empire. The Lycurgus Cup is a "cage" cup, a work of art constructed from a single glass carved with utter care to depict a relief of the scourge of Lycurgus at the hand of Dionysus just fractions of centimeters away from the exterior of the drinking vessel itself. Lycurgus is a key figure in Homer's *Iliad*, and, according to tradition, the king of Edoni banned the cult of the god Dionysius (whose Roman equivalent is the much better known Bacchus). Lycurgus did not stop there, however. He proceeded to remove any traces of the god of wine along with his followers from the realm of Edoni.

While the carving is interesting—alone worthy of artistic esteem when one considers the fragility of the material—the true beauty is in how the cup reflects light. When lit from within the bowl-like opening, the Lycurgus is a dull green; however, when light reflects off the relief, the cup takes on a blood-red appearance. The delicate play of light and color bouncing off the Lycurgus Cup gives the relic a mystical effect, an effect no doubt heightened due to the cultural atmosphere of the period.

The Lycurgus Cup has been damaged over the seventeen hundred years that have passed since its near-miraculous creation, with a silver overlay affixed to the rim of the cup and along the stem to ensure its structural integrity.[3] Even without the color change effect made possible by the addition of ground-up gold in the block of glass the cup was hewn from, the Lycurgus Cup would still be a cherished historical relic due to the type of carving and possible religious ties.

After fifteen hundred years of anonymity, the Lycurgus Cup resurfaced in 1845. The cup changed hands several times over the next century, eventually becoming a part of the expansive art collection of Great Britain's Rothschild family. Lord Victor

Rothschild negotiated the acquisition of the cup by the British Museum in 1958, where the remarkable feat of ancient craftsmanship resides today. Although the cut-glass cup is quite fragile, the British Museum has taken steps to study it, enlisting General Electric to analyze a fragment from the cup. The General Electric scientists found traces of gold and silver in the glass, spurring discussion as to how and why the metals were placed in the glass. It is quite likely that the metals were intentionally ground and placed within the melted silica that later formed the block of glass from which the Lycurgus Cup was carved.[4]

We do not know the exact method by which the Lycurgus Cup was crafted, but we can make some educated conclusions as to why tiny pieces of ground gold and silver added in careful proportion yield a color-changing effect. If the individual pieces of metal added to the molten-glass mixture are small enough—in the low nanometer range—the metal begins to take on interesting properties. These tiny pieces of metal are then classified as a new form of particle—nanoparticles—with their incredibly small sizes conferring new and strange characteristics to the metals, properties completely different from those exhibited by the same metal when found in its natural state.

Most nanoparticle mixtures are in the ten- to one-hundred-nanometer range—such a small size that it is quite difficult to have an inherent frame of reference to imagine the tiny pieces of metal. A million nanoparticles lined in a row would barely make it from one end to the other of your television remote control's power button, while a US one-dollar bill is a little less than eleven thousand nanoparticles thick when fresh off the press.

The behavior of nanoparticles in the path of light can change with small increases or decreases in size. The metallic yellow color of a gold bar is universally known; however, a colloid of gold (a mixture of gold in a solvent, usually water) nanoparticles thirty nanometers in diameter is a reddish-orange color. Increase the diameter of those nanoparticles slightly, and the gold colloid is now purple.[5] A drastic change in color, but one made possible due to the radically different properties of nanoparticles.

Both the gold and silver in the Lycurgus cup are subject to the same change in properties seen with gold colloids, with another property familiar to physics students coming into play—dichroism, the ability for a material to reflect two different wavelengths of light simultaneously.[6] In the Lycurgus cup, the altered color properties of gold and silver nanoparticles combine with dichroism to result in the viewer seeing a change in color from pale green to wine when the light source changes position.

PROVING THEIR DISCOVERY

What set Antonio de Ulloa and Julius Caesar Scaliger apart in the discovery of platinum is their mention of the metal in texts each wrote later in life. What may be more telling, intellectually, is information pointing us in the direction of those who championed the use of platinum by manipulating its uniquely stable characteristics. British scientist Charles Wood smuggled a small cache of platinum from South America in 1741 and distributed the metal among colleagues in Europe, proselytizing its characteristics with so much vigor that the scientific community of his day added platinum as the eighth known metal, making it the first added since ancient times (iron, gold, silver, tin, mercury, lead, and copper round out this primitive roll).[7]

Those in scientific circles nicknamed the metal "white gold," a term now co-opted to describe rhodium-plated gold jewelry. Following up on the success of Charles Wood, a pair of business-minded scientists, William Hyde Wollaston and Smithson Tennant, initiated platinum export efforts a little over six decades later. The duo exploited the most steadfast characteristic of the metal—its stability in harsh conditions—to make laboratory equipment that would not tarnish, and in doing so began the mainstream use of platinum.[8] The ability of a large piece of platinum to remain inert—an overall resistance to chemical reactions—led France to painstakingly manufacture a platinum bar of exactly one meter in length to use as the universal standard of length in the metric system.[9]

Today platinum does not enjoy the allure and traditional value that accompanies gold, with this disconnect forcing platinum to be more susceptible to price fluctuations than its precious-metal forefathers. This gap is becoming smaller with each passing year as we find new ways to take advantage of the properties of platinum. At the rate of change in our valuation of *platina del Pinto* over the past three centuries, our progeny may very well hold platinum in higher esteem than gold.

CHAPTER 14

THE NEXT PRECIOUS METALS

Humankind's history of mining and cherishing gold reaches back to about 4,600 BCE, with Bulgaria's Varna Necropolis showing that our ancestors valued the soft yellow metal as much as we do, if not more. Three hundred and ten bodies are buried at Varna, but the mass burial is not what makes the site remarkable. Enclosed in a number of the graves are gold trinkets, making this site one of the oldest known examples of humans bestowing on gold a value higher than on iron, lead, or other metals used during this period.[1] The gold items vary in size (one of the more outlandish treasures was a golden penis shaft), utility, and placement. The artifacts are distributed evenly at Varna, with the golden trinkets likely noting the lofty status of those buried alongside.[2]

As time passes, the materials society covets change, in part because human ingenuity and technology meet to create unimaginable needs. What follows is a look at a handful of rare and scarce metals with rapidly evolving applications. While they may not achieve the historical esteem of gold or silver, humankind will no doubt value many of them more at the end of the twenty-first century than they do at the moment. Which one of these metals will become the element we just cannot live without?

THE DEPARTMENT OF DEFENSE'S LUST FOR BERYLLIUM

The United States cannot produce useful quantities of eight of the seventeen elements commonly labeled as rare earth metals—

terbium, dysprosium, holmium, europium, erbium, thulium, ytterbium, and lutetium—because they simply do not exist within our borders. This situation, one often brought to light by individuals making a case for building rare earth metal caches in the name of national security, makes it necessary for those seeking the materials to look abroad, an act often complicated by political relations regardless of the extent of the purchaser's association with the home government.[3]

Governments focusing on a specific substance for acquisition is not unusual, although we are more accustomed to struggles over consumer materials like oil or, in dire cases, foodstuffs such as wheat or rice. But governments have targeted specific elements in the past. According to the US Department of Defense, high-purity beryllium is necessary to "support the defense needs of the United States during a protracted conflict," but procuring a supply is not easy.[4]

Making a case for the defense industry's reliance on beryllium is easy. No fewer than five US fighter craft, including the F-35 Joint Strike Fighter that will be employed by the United States, Japan, Israel, Italy, and five other countries over the next several decades, rely on beryllium to decrease the mass of their frames in order to allow the nimble movements that make the planes even more deadly. Copper-beryllium alloys are a crucial component of electrical systems within manned craft and drones, along with x-ray and radar equipment used to identify bombs, guided missiles, and improvised explosive devices (IEDs). The metal also has a use far removed from such high-tech applications. Mirrors are fashioned out of beryllium and used in the visual and optical systems of tanks because it makes the mirrors resistant to vibrational distortion.

High-purity beryllium is worth just under half a million dollars per ton when produced domestically, with Kazakhstan and Germany supplying the only significant amounts to the United States through import.[5] In 2008 the Department of Defense approved the construction of a high-purity beryllium production plant in Ohio after coming to the conclusion that commercial domestic manufacturers could not supply enough of the processed metal for defense applications nor did sufficient foreign suppliers exist. While the plant in Ohio is owned by a private corporation, Materion, the Department of Defense is apportioned two-thirds of the plant's annual output.[6]

LANTHANUM AND THE ELECTRIC CAR

Lanthanum is the namesake of the lanthanide series, the section of the periodic table that holds fifteen of the seventeen rare earth elements. Lanthanum hasn't had many everyday applications—it's been used in welding equipment, in lamps used for camping, and in movie theater projectors—but the metal's status has increased dramatically with widespread consumer acceptance of the electric car. Lanthanum is the key component of nickel-metal hydride, with each Toyota Prius on the road requiring twenty pounds of lanthanum in addition to two pounds of neodymium.[7] Like many of the rare earth metals, lanthanum is not as rare as the description would suggest; it is the separation and extraction of lanthanum that complicates matters and thereby results in the metal's relative scarcity. With the Nissan Leaf and Tesla Motors' Roadster becoming trendy choices for new car buyers, the need for lanthanum will remain and no doubt grow in the foreseeable future. The metal will become even more relevant as automobile manufacturers push the limits of battery storage, an effort that will require significantly more lanthanum for each car rolling off the assembly line.

GRAPHENE

First off, I should clear the air—graphene is not a metal. The material is a sheet of repeating carbon atoms oriented in an array identical to that of a bee's honeycomb, but the material exhibits a great many physical characteristics that are prized in metals. A layer of graphene is extremely thin—the width of a carbon atom—but due to the number of unusually oriented carbons in the molecule, the material is stronger than steel and comparable to that of diamonds.[8] The material also conducts energy very well, a fact that puzzled scientists who explored the upper limit of graphene's ability to conduct heat.[9] It became a darling of chemists, physicists, and engineers shortly after its discovery in 2004, a discovery for which its creators—Russian scientists Andre Geim and Konstantin Novoselov—won the 2010 Nobel Prize in Physics.[10] Using graphene to fill existing needs seemed to be the work of science fiction at the time of its discovery, but an outpouring of interest, effort, and funding from academic and corporate institutions may place graphene-containing products on the shelves of stores within the next decade. Creating large amounts of high-quality graphene was the major stumbling block until early 2014, when Samsung's Korea-based Advanced Institute of Technology published in the journal *Science* a promising synthesis method using the metalloid germanium to make uniform, and thus usable, sheets of graphene.[11]

Foreseen applications of graphene already run the gamut of imagination and include aircraft bodies, large-scale water desalination, super-efficient transistors, radioactive waste disposal, and even condoms.[12] Yes, condoms. In 2013, the Bill & Melinda Gates Foundation awarded a one-hundred-thousand-dollar grant to create ultrathin condoms out of graphene in the hope the high-tech prophylactics would be of aid to the developing world.[13] Manufacturers of three-dimensional printers are eagerly awaiting widespread availability of graphene; a printable form of the molecule would allow for the made-to-order creation of strong, lightweight, and imaginative objects at a moment's notice.[14] The ability to efficiently synthesize graphene will remove any aspect of the secondary impacts on financial markets seen with gold, platinum, and other desirable metals. Large-scale synthesis of graphene makes the material worthless (no longer rare) and yet invaluable at the same time, with increased availability making graphene a material with so many possible uses that we might not be able to live without it.

THORIUM AS A REVOLUTIONARY ENERGY SOURCE

The metal thorium is often viewed as the bane of many mining operations because it is a radioactive metal found in substantial quantities in rare earth–bearing minerals and lacks substantial value due to having few modern applications. Before society became aware of the dangers of close-quarters radiation exposure, thorium found use in handheld gas lamps and as an additive in glass manufacturing.[15]

Thorium components within lamp mantles have long been replaced with the rare earth metal yttrium; however, an old leftover mantle remaining on store shelves may contain thorium. Long-term exposure to thorium mantles (say, sitting by a campfire every weekend for decades) will probably not bring any harm; however, those exposed to a large number of the fragile lamp mantles on a daily basis—individuals working in a manufacturing capacity—tread a dangerous line. Inhalation of thorium-containing molecules in a factory environment is extremely difficult to protect against, leaving the workforce subject to thorium exposure and deposits lodging in their lungs.

The interior of lamp mantles, however, are a much safer hiding place for Thor's metal than the twentieth-century use of the element as a toothpaste additive. The German corporation Auergesellschaft transformed leftover thorium from their lamp-mantle manufacturing business into the primary selling point of *Doramad Radioaktive Zahncreme*. Auergesellschaft touted the bacteria-killing benefits of radioactive thorium on the package, while conveniently overlooking the downside of radiation exposure.[16]

The future uses of thorium to generate light and energy were far from the minds of Swedish chemist Jöns Jakob Berzelius and Norwegian mineralogist Reverend Morten Thrane Esmark when the pair codiscovered the element in 1828 and christened the metal with a glancing nod to the Norse god of thunder. Two centuries later, however, the name and the images it conjures appear apt. While thorium is radioactive, it is far less dangerous than uranium, and a new generation of chemists, physicists, and nuclear engineers believe thorium is the key to a new, safer source of nuclear energy.

A thorium nuclear power plant would operate in a radically different manner than the uranium solid fuel plants we are accustomed to supplying our electrical needs. Solid fuel plants rely on uranium beads embedded in shielded rods and immersed in water. The water is kept under pressures over a hundred times the normal atmospheric pressure so the liquid can continue to mediate heat from the solid fuel without evaporating. The first solid fuel nuclear power plant in the United States opened in 1958 in the small town of Shippingport, Pennsylvania, as part of President Dwight Eisenhower's "Atoms for Peace" initiative. The plant operated for twenty-five years with minimal incidents before being mothballed.

Despite the solid track record of most nuclear power plants, when these plants do fail, the results are disastrous. When a solid fuel nuclear power plant fails to control the energy released from countless nuclear fission reactions occurring simultaneously, the core undergoes a meltdown. The high-pressure liquid water surrounding the uranium fuel evaporates, and the now-unstoppable nuclear reaction melts the solid uranium beads. A common belief during the fearful dawn of commercial nuclear power plants entailed the now liquid nuclear fuel continuing to melt, piercing the ground and continuing to push until the fuel reached the other side of the planet. This idea was popularized in the 1979 film *The China Syndrome.*

Two other major types of nuclear reactors exist—liquid fuel and molten salt—but neither see widespread use. In liquid fuel reactors, energetic uranium compounds are mixed directly with water, with no separation between nuclear fuel and coolant. Liquid fuel reactors can make use of lesser-quality uranium and appear to be safer at first glance because the plants do not need to operate under high pressure to prevent water from evaporating. On the downside, they pose an even larger contamination and waste storage problem than conventional solid fuel reactors. Since there is no separation between the cooling waters and uranium, much more waste is produced, waste that, in theory, must be stored for tens of thousands of years in geological repositories before the murky waters no longer pose a danger.

The second type of nonconventional nuclear reactor, the molten salt reactor, is the subject of renewed interest and could make use of the excess thorium found in the process of rare earth mining.[17]

A radioactive thorium-fluoride salt would be used to power these plants, along with a small amount of uranium used to initiate the reaction. Any radiation emitted by the starting sample of uranium would breed more atoms of uranium by forcing the addition of a neutron to thorium-232, which decays into the energy-generating uranium-233.

Dreams of a thorium reactor are not far-fetched—the United States operated a full-scale thorium reactor for power generation within the military boomtown of Oak Ridge, Tennessee, between 1965 and 1969.[18] Oak Ridge is no stranger to secret military projects, being the cornerstone of uranium enrichment efforts during the Manhattan Project. During the 1950s and early 1960s, thoughts of thorium-powered homes were not far from the minds of government physicists. The United States prototyped thorium reactors for use in bomber jets in 1954, with hopes that thorium could provide a virtually unlimited source of power to maintain vigilance in North American and European skies against nuclear threats during the Cold War.[19]

One of the promising aspects of a molten salt reactor is that the system is, at least in theory, meltdown-proof. In the event of an electrical failure that would prevent the team of engineers from controlling energy release, a refrigerated plug in the vat holding the molten salt mixture would cease to be cooled and the heat of the reactions ongoing would melt it. Once the plug is dissolved, the molten salt will flow into a specially designed containment unit, ending a runaway cascade of energy generation and saving billions of dollars, the reactor facility, and lives in the process.[20]

Thorium reactors also solve, or at least smooth over, the awkward junction between national security matters and nuclear power. If the thorium salt used as a fuel by these reactors is stolen by a terrorist group or foreign nation for use in a nuclear weapon, the culprits are in for a surprise. The thorium mixture used inherently gives off an easily detectable isotope of the rare metal thallium. This isotope, thallium-208, is created from decaying uranium-232. Thallium-208 releases an intense stream of gamma radiation, one that would threaten the lives of those working with it in a matter of moments. The bomb would need heavy shielding at all times in the manufacture process to prevent government entities from detecting the radiation signature (the United States maintains the Integrated Operational Nuclear Detection System explicitly to detect this type of threat), with the bomb giving off a substantial "ping" once placed in position.[21]

If the capability to create safe thorium power plants is several decades old, why do all commercial nuclear reactions make use of solid fuel uranium metal to generate energy? No truly satisfying answer to this question exists. Rumors of nuclear power funding decisions suggest President Richard Nixon eschewed commissioning commercial molten salt reactors in favor of solid fuel reactors that would be researched and developed in Nixon's home state of California in order to bring economic benefits to the area.[22]

Another line of thought posits that thorium reactors were overlooked because the fuel did not create weapons-grade waste as a by-product, a perceived benefit of solid fuel uranium reactors.[23] Both of these ideas appear plausible—bureaucratic stumbling blocks are a part of everyday political life, while research endeavors and the military are familiar partners. The lack of legitimate facts is bothersome, dimming the lights on both of these theories.

A third reason, however, appears more plausible due to its pleas for practicality. Thorium power plants would need constant maintenance and a highly skilled set of workers on around-the-clock watch to oversee energy production. This is not to say solid fuel nuclear power plants are worry-free, but the solid fuel plant is the comfortable dinner-and-a-movie alternative to taking a high-maintenance individual out for a night on the town. Why would molten salt plants need constant observation? Thorium molten salt reactors create poisonous xenon gas, a contaminant that must be monitored and removed to maintain safe and efficient energy generation.

Because of this toxic by-product, a thorium molten salt reactor would not succeed with just a technician overseeing a thoroughly automated plant but would require a squad of highly educated and dedicated engineers analyzing data and making changes around the clock. Luckily, most of the world's current power plant employees are quite educated, but the act of retraining each and every worker is a substantial barrier that prevents the switch to thorium fuel plants in North America. The situation might change in a few decades as the bevy of solid fuel plants built in the United States during the 1970s and '80s reach the age of decommissioning during the middle of this century, but for now, putting into place thorium nuclear technology is best left to countries still climbing the ladder of nuclear power.

Currently, the world's two most populated countries are ramping up efforts to introduce thorium-based nuclear energy. China is unperturbed by the intellectual hurdles accompanying large-scale production of electricity via thorium nuclear reactors, with the country aiming to increase the four-hundred-plus scientists and engineers currently employed in thorium electricity generation to seven hundred and fifty by sometime this year. No country currently possesses a functional thorium plant, but China is on the inside track thanks to an aggressive strategy that aims to begin electricity generation by the second half of this decade.[24] India is committed to generating energy using thorium as well, aiming to make use of their own extensive thorium reserves to meet 30 percent of their energy needs by 2050.[25]

NEODYMIUM AND NIOBIUM

Few childhood playtime memories top the ones I have of a simple magnet and a huge hunk of metal. The gentle tug of a refrigerator magnet while drawing closer to a metal light pole, the whimsical ability to manipulate tiny objects by merely moving a magnet nearby. Finding a magnet of sufficient quality and size was difficult a couple of decades ago, but nowhere near as problematic as the present because child safety requirements have all but removed the magnet as a scientific toy for fear of damaged fingers and pinched intestines should a child manage to swallow their play toy.

If you've ever entertained yourself with an errant refrigerator magnet waiting for the oven to preheat or dug through a toolbox for one to magnetize a screwdriver as you whittle away at a household "to do" list, the strength of neodymium magnets, often ten or more times stronger than a run-of-the-mill refrigerator magnet, might catch you off guard. Neodymium—one of the two elements derived from Carl Gustaf Mosander's incorrect, but accepted, discovery of didymium in 1841—is the most widely used permanent magnet, with the rare earth metal being found in hard drives and wind turbines as well as in lower-tech conveniences like the button clasp of a purse. Along with the rare earth metal neodymium, niobium metal magnets are becoming increasingly necessary in recreational items, in particular, safety implements, electronics, and the tiny speakers contained in the three-hundred-dollar pair of headphones your child wants for his birthday.

Niobum receives its name through a quirk of Greek mythology. Traditionally, Niobe is known as the daughter of Tantalus (for whom the rare metal element tantalum is named). Like her father, she is a thorn in the side of the gods. Niobe is lucky in one part of life—she is the mother of fifty boys and fifty girls, and she takes a considerable amount of pride from this fact. Her pride is too much for Apollo and Artemis to take—the mythological super couple are only able to bear a single boy and girl, and when Niobe gloats in their midst, Apollo and Artemis slay all one hundred of Niobe's offspring. Mass murder is not enough to quench the godly anger in this bummer of a story, as Apollo and Artemis take the scenario one step further and turn Niobe into stone.

Niobium, a metal typically used to make extremely strong magnets, is also quite stable and has the added bonus of mild hypoallergenic properties—a boon to the medical world in which niobium became an obvious choice for use in implantable devices, specifically pacemakers.[26] The selection of complementary inert metals in biological implants is an important engineering task since the last complication a physician or patient would want is the rejection or inadvertent breakdown of the structural components after a successful transplant.

Permanent magnets, as the name suggests, do not lose their magnetic properties. They are a reliable but more expensive alternative to electromagnets, which use an electric current and coiled metal wire to create a magnetic field. Magnetism and electricity go hand in hand in modern life—magnetic fields affect electrical fields and vice versa. This connection is used to create superconducting magnets, which run electrical current through metal coils to generate the strongest magnetic fields possible with our current understanding of technology. Using a wire made of a permanent magnet, like neodymium, turns the basic run-of-the-mill electromagnet into a superconducting one.

Superconducting magnets are at the heart of investigations to discover new elements. How you ask? These magnets are aligned, often hundreds at a time, to create the rings used in particle accelerators. Thanks to the pulling and pushing we all attribute to magnetism, tiny particles—often individual neutrons or atoms—are propelled at unbelievable velocities at targets on the other side of the ring; scientists investigate the aftermath of these spectacular crashes to look for evidence of the creation of new elements. The European Organization for Nuclear Research's (CERN) Large Hadron Collider used advanced magnet technology to the search for the Higgs particle, with the LHC's superconductors built with magnets made from niobium-titanium alloys.

China and Japan apply technology of this type to propel humans across their respective countries at two hundred and fifty miles an hour in maglev trains at land speeds seen only in Formula 1 races and at four times the speed of a New York City subway car. Lacking wheels, these trains float millimeters above a specially designed surface and are propelled by the electrical manipulation of magnetic fields.

A derivative of magnetic rail technology may be used to one day send cargo to space. The US Navy is currently developing magnetic field–powered rail guns that do not need expensive ammunition or a chemical propellant. Instead, they use superconducting magnets to propel inexpensive, thirty-pound aluminum blocks to devastate enemy armor. This technology could be used for a kinder purpose, namely, a less expensive manner of launching cargo and, with a number of safety measures in place, people into space. In theory, magnetic field launch–systems would use superconducting magnets to propel pods of cargo on a banked track at speeds high enough to allow the cargo pod to escape the earth's gravitation pull. These launch systems need extremely long tracks—several miles long in some cases—to give the pod enough time to reach escape velocity with reasonable energy expenditure. Making use of short launch tracks would be possible, but the rate of acceleration for the projectile would need to be increased dramatically.

Estimates claim a magnetic launch system would slash the cost of launching payload into orbit by 95 percent over current chemical propulsion methods.[27] Richard Garriott's 2008 flight into space aboard the Russian Soyuz to the International Space Station cost the video game entrepreneur (and, as you will learn later in the book, stranded spacecraft owner) thirty million dollars.[28] Media reports painted the revenues of this episode of space tourism as pure profit, but a large chunk of the thirty-million-dollar ticket price went to paying the cost of sending an extra passenger into space. Current NASA figures place the cost of sending cargo into space at ten thousand dollars per pound, but magnetic-launch systems cost less than 1 percent of that amount.[29]

This financial factor is what will lead corporations and governments to turn to magnetic field–launch systems if the technology is eventually perfected. Tracks could be placed on every continent and could meet all payload launch plans, with countries leasing time on multiple tracks if needed. This form of low-cost space transport—one that makes use of rare metals—could also have a major impact on the future of how humans acquire rare and scarce metals, a possibility we will visit later.[30]

Transportation and applied physics research are not the only sectors that could be revolutionized in the next few decades. When molecular biologists and biochemists examine cell cultures to study the growth of bacteria or diseases like AIDS, they grow a number of these cultures to act as their biological playing field. At the moment, bacteria cultures are grown on two-dimensional petri dishes. Petri dishes are inexpensive, although observing growth on a two-dimensional platform limits our ability to see how cells and viruses truly behave as living organisms due to a simple but important fact—life takes place in three dimensions. By coating cells with magnetic nanoparticles and placing them in the presence of a small but powerful neodymium magnet, the experimental cells are levitated within a nutrient supply.[31]

In the artificial bacterial field, cells begin to interact with one another during growth and replication, forming associations more closely resembling tissue and interactions observed within humans and other mammals than a two-dimensional cell growth allows. Cell-to-cell interactions in living tissue naturally take place in three dimensions, and three-dimensional cell culture technology made possible by neodymium magnets will allow scientists to better assess the behavior of vaccines and pharmaceuticals prior to in vivo trials.[32] The use of three-dimensional cell cultures is still in its infancy, but we could very well see the widespread use of these cultures in the day-to-day operations of academic and industrial laboratories across the world by the end of the next decade.

COULD HAFNIUM ALTER THE FUTURE OF WARFARE?

Hafnium is a rarely used metal, but the seventy-second element on the periodic table could change the future of warfare, thanks to the outcome of a handful of experiments conducted in a small Dallas, Texas, lab. The experiments involved exposing a minute amount of a nuclear isomer of hafnium-178 to a beam of x-rays. Nuclear isomers differ from the typical definition of isomer in that the number of neutrons is stable, but one or more of the neutrons carries with it an inordinate amount of energy. With the hafnium-178 placed on a disposable Styrofoam coffee cup, researchers exposed the metal sample to a beam of x-rays, the same kind one would be exposed to during a routine dental examination. In fact, C. B. Collins, the principal investigator on the project, used a commercial dental x-ray source to perform the experiments.[33] This is a far cry from the heavily financed beginnings of the cyclotron or CERN's Large Hadron Collider, but science favors those prepared to acknowledge an unusual result, regardless of the mechanism by which the result appears or the resources available.

In ten minutes of data collection spread out over a twenty-four-hour period, Collins and his team witnessed a phenomenal event. When they placed the target isotope of hafnium in the path of an x-ray, the researchers observed the release of a remarkable amount of energy from the sample. Could the impact of x-rays actually be triggering a release of latent energy seeded within each atom? Collins reported that the energy released exited as gamma rays—gamma rays and x-rays are simply forms of energy—and appeared to have shown that the long-speculated possibility of induced gamma emission was now a reality.

Collins's report drew controversy from the first day of publication, with skeptics puzzled as to why the groups did not observe a much-needed increase in decay rate while in the presence of x-rays—a corollary physicists expected to witness. Naysayers also voiced displeasure over the low resolution of the data, suggesting that the energy release, much like the detection of masurium by Ida Noddack, was manufactured in the minds of scientists looking too deeply into ever-present experimental noise.

Energy released through triggering this reaction in excited samples would be hundreds of times larger than the amount of energy released through the detonation of a stick of TNT, but considerably less than the potential released when a nuclear bomb explodes. This is a sweet spot for many energy projects and viable if it becomes not only a controllable range useful in a number of opportunities but also the bleeding edge of environmentally friendly alternatives to coal and natural gas.

Such an enormous energy output would no doubt draw the attention of the military-industrial complex as a tool more powerful than current field munitions but without the international controversy and environmental fallout following detonation of an atomic weapon. Naturally, when government officials learned of Collins's hafnium experiment some of the best minds and resources were tasked with picking apart flaws in the work and determining the real-world feasibility of its use.

While the energy payoff claimed by Collins appeared extraordinary and maybe even a bit unreasonable, going ahead with research into the phenomenon was a noble effort. Groups around the world began to show interest in hafnium research, including a handful of scientists working inside government-funded Russian labs. For the United States, waiting any amount of time might have led to a net loss in footing against world powers, and for the bureaucrats allotting budget money, the end of their career, should another entity gain success with hafnium projects that they rejected.

The US government previously tasked the JASON Defense Advisory Panel, a small group of scientists, with providing cutting-edge opinions while weighing and imagining the dangers present with this advancing technology. The background of the JASON group feels like a tale ripped from a spy novel. Born out of the early days of the Cold War and financed through the MITRE Corporation, the nonprofit think tank has long-standing ties to the United States military. Members of the group meet in an undisclosed location during the summer months to discuss topics that may impact the future of not only the United States but the world. In the past, the JASON group weighed in on low-cost DNA sequencing, manufacturing petroleum replacements from bacteria, the impact of solar flares on the planet's electric infrastructure, improving human mental performance by surgically implanting processors into the brain, and the future of terrorist attacks. Their discussions and findings are summarized and a version of their report is made accessible to the public at large.[34]

The identities of JASON members are not formally disclosed, but one can assemble a partial roll from the documentation page attached to their published studies. New additions to the JASON group are hand-selected by current members. This practice raised the ire of the group's longtime sponsor, Defense Advanced Research Projects Agency (DARPA), which eventually cut ties with JASON in 2002 over their refusal to alter the selection process.[35]

When the JASON group gathered to discuss the viability of hafnium explosives as a military weapon during a meeting funded by DARPA, the verdict was quite critical due to the lack of experimental evidence at the time of the meeting in 1997.[36]

Even if the United States failed to unleash a hafnium "bomb" during the course of battle, the research would provide invaluable insight into the ability of other countries or clandestine individuals to make use of the technology. If such a weapon could be made, it is best to know how it works, and if not, the discussion is over and no military leader loses another night's sleep mulling over the destruction a hafnium bomb might create. Passing over a hafnium weapon would also be career suicide for any military or government official should the weapon prove viable, putting senior officials in a position to give the go-ahead for projects they may not believe in—the possible payoff is too high to pass up. Being hit by a beam of x-rays causes the energy stored within the individual atoms to be released in a quick flood. This energy would typically be released in trickles of gamma rays over the thirty-one-year half-life of hafnium through normal decay, with the possibility of an all-at-once release making for an ideal weapon when one is looking for an explosive with a higher yield than conventional detonation but that is shy of a nuclear detonation. Extrapolations from Collins's research suggest the excitement of a single ounce of pure hafnium would release enough energy to boil one hundred and twenty tons of room-temperature water.[37]

What makes hafnium immanently weaponizable is the small amount of fuel—the hafnium sample itself—and the relatively simple and prevalent stimulator needed—portable x-ray machines. The inner workings of a hafnium weapon provided fodder for another debate, this time a legal one. Proponents of hafnium research argued that a weaponized hafnium armament would not violate the Comprehensive Nuclear-Test-Ban Treaty, allowing the United States to make use of such a weapon while skirting the United Nations and its pact with member countries. Whether or not a hafnium bomb is nuclear energy "release" or "explosion" is subjective in nature, but the bomb no doubt violates the spirit of the treaty.[38]

Academic research findings, unless covered by patent implications or secrecy agreements, are traditionally made public through scholarly articles. These articles are the currency of the ivory tower, with professorships earned, grants awarded, and careers sent careening into mediocrity every day based on one's ability to produce a steady stream of research, let alone a short four-page article that could have an impact in the real world. The public nature of scholarly research plagues future reports of similar hafnium experiments—an academic scientist's desire to publish information is juxtaposed with a tight seal on research involving a superweapon. This flies in the face of the science, but, at the same time, it is a stance that protects both the citizens of a country and their investment. In addition to hafnium, nuclear isomers of the coveted metals tantalum, osmium, and platinum exist that are capable of undergoing induced gamma emission.

The Pentagon has long sought a nuclear-powered unmanned aerial vehicle, a vehicle capable of extending drone missions lasting from an afternoon to a hundred-plus days. In 2003, the US Air Force Research Laboratory explored the possibility of using the gamma-inducing characteristics of hafnium-178m2 to reach the same result, but with no success.[39]

CHAPTER 15

WHEN THE WELL RUNS DRY

The natural resources of Antarctica may one day be counted and distributed throughout the world, but deciding who will benefit from the treasures of the frozen tundra is a conundrum. Antarctica lacks, as Pulitzer-winner Jared Diamond eloquently states, "a self-supporting population" in his book *Guns, Germs, and Steel*, with its only residents temporary and citizens of political and economic powerhouses. With no people of its own and, therefore, no government, Antarctica is in a prime situation to become the treasure chest of developed countries across the world.[1]

The first modern agreement governing the investigation of the seventh continent was signed in 1961. The Antarctic Treaty banned military activity on its landmass—a serious geopolitical concern during the Cold War—and established a pro-science atmosphere that would bring eminent scientists from across the world to observe its harsh environment. Measures governing the resources of Antarctica would be reinforced in 1991 with the Greenpeace-championed Madrid Protocol, which sought to eliminate problems stemming from waste buildup on the scarcely inhabited continent and to prevent exploration or mining for financial gain. The Madrid Protocol further establishes Antarctica as the world's foremost nature reserve and prevents any signing nation from mining or otherwise acquiring a stockpile of natural resources from the continent. As it stands, there is a definite risk that the treaty will be abandoned or heavily amended when the protocol comes up for renewal in 2048.[2]

If the opportunity arises, how will the nations of the world parcel out the resources of Antarctica? Since it currently lacks a formal government, the land would likely be divided between the major powers on the planet in 2048, or funds would be provided to a central financial body entrusted with overseeing the resources of the continent. The latter was an option provided in a failed 1988 treaty proposed by New Zealand: the Convention on the Regulation of Antarctic Mineral Resource Activities, which would have allowed for the mining of Antarctica as long as permission was acquired from a third-party international clearing house.[3] Thirteen countries—the United Kingdom, New Zealand, France, Norway, Australia, Norway, Chile, Argentina, Brazil, Peru, Russia, South Africa, and the United States—currently lay claim to areas of Antarctica or reserve the right to claim a swath of its land in the future under the original provisions of the 1961 Antarctic Treaty. Interestingly, the United States and Russia are two of the countries yet to officially announce a claim.[4] While these demarcation lines are not universally accepted, each of these thirteen countries also signed the 1991 Madrid Protocol, lending a degree of formality to the previous claims.[5]

Australia currently holds the largest claims with respect to Antarctica, which stem from the country's expeditions of the continent at the dawn of the twentieth century. The Australian Antarctic Territory encompasses nearly six million square kilometers and 42 percent of the entire continent, turning the investment made in a series of early expeditions into a resource boon should the continent open up for mining operations and Australia hold onto its perceived claim.[6] Although mining in Antarctica is banned, observation and assessment of available natural resource supplies in the name of science is allowed and encouraged. Over fifty international research stations are spread across the continent, with four thousand scientists residing in-house, including three thousand sent by the United States.[7] While sheets of ice several hundred feet thick cover most of the surface of Antarctica, scientific expeditions have identified deposits of natural gas, petroleum, and coal reserves as well as gold and even diamonds.[8] Rare earth metals have been identified after analysis of cylinders of ice drilled from the ice sheets covering Antarctica—the only thing standing in the way is the go-ahead for mining.[9]

PILLAGING GREENLAND

Although mining of Antarctica is at a standstill, recent political developments could turn an existing country into a training ground for operations in Antarctica. In 2013, the parliament of Greenland—the rocky, gray landmass next to the almost comically lush island of Iceland—voted to lift a two-and-a-half-decade ban on the mining of radioactive materials in the country, making real the possibility of mining for a number of rare earth metals within the country's borders.[10]

Straddling the Arctic Circle, Greenland is at the same time the largest island on the globe and the least densely populated country in the world, serving as home to just shy of sixty thousand inhabitants in 2013. If the entire population of Greenland crammed itself into its capital city of Nuuk, the population of the now-swollen city of Nuuk would be on par with the bustling metropolis of Manhattan, Kansas.

What makes Greenland's decision interesting is the possibility of finding rare earth metals. Greenland has long been hypothesized to have rich resources of the metals, but any and all attempts at commercial mining have been halted because uranium is commonly discovered during excavations of rare earth metals. Once Greenland's parliament overturned legislation banning the extraction of uranium, the parliament also freed up the country for mining of treasure troves of rare earth metals.

Even the formerly forbidden uranium frequently found in trace amounts with rare earth metals could be viable for use, leading to another revenue stream for the country. Selling uranium is a far trickier feat than selling forty or fifty tons of neodymium and praseodymium containing monazite ore. Due to safety concerns, the handling and sale of uranium by-products will likely be controlled by the Kingdom of Denmark, which continues to provide armed forces to maintain the security of Greenland.[11]

In a lot of ways, the icy underbelly of mining in Greenland is a microcosm of what nations and corporations will face if the embargo on resource acquisition in Antarctica is lifted in the latter half of the twenty-first century. Along with rare earth metals, raw natural gas and oil resources are present, mirroring the assets of Antarctica. In addition, nearly 80 percent of Greenland is covered in an ice sheet ranging in thickness from a few meters to two kilometers, with the island receiving enough seasonal temperature change to cause the sheet to recede along the edges. Technology used to drill and mine in Greenland will be extremely useful if and when Antarctica is opened to the nations of the world.

The abysmal economy of Greenland would welcome the establishment of a thriving industrial mining community; at present it remains viable only through economic handouts from the Kingdom of Denmark. In 2008, the citizens of Greenland successfully passed a referendum for increased autonomy from the Kingdom of Denmark, but an infinitely small population—65,360 inhabitants in 2013—spread across the twelfth-largest country by area in the world will forever pose a stumbling block.[12] Less than half of the country's population are classified as active members of the labor force, with unemployment hovering in the 9 to 10 percent range.[13] If the people of Greenland truly want the chance to stand on their own, mining their uranium and rare metal resources would be a wise choice. Early estimates suggest that Greenland's supply is sufficient to supply a quarter of the planet's needs for the next five decades—an easily exhaustible amount, but at the same time a wealth of resources worthy of exploration if demand warrants.[14]

Ideally, operations in Greenland will be free of the current and continuing political restrictions and paranoia that plague mining in China. Whether Greenland's existing ties to Denmark and the European community will continue is a point of contention, as is the possibility of establishing preferences when Greenland's Bureau of Minerals and Petroleum doles out territory rights to governments and corporations. If managed correctly, Greenland could position itself as a rare earths leader in the Western Hemisphere as China has cunningly done in the East.

PURIFYING MUD

After Greenland and Antarctica, the options for new sources of rare and scarce metals become increasingly uncertain. First, there's mud—"red mud," a waste product from mining spread across the nation of Jamaica, home to a number of industrial bauxite mining operations. Bauxite is a mineral mined as an aluminum resource. Surveys performed by Jamaica's all-encompassing Ministry of Science, Technology, Energy, and Mining show that this red mud still has value, with a fair amount of rare earth metals dissolved in the waste product.[15]

Divining the origin of red mud is not as easy as its name would suggest. Red mud is not actually mud but is rather a slurry of metals and particulates leftover from industrial-scale refining of bauxite en route to obtaining aluminum. Red mud is created through the Bayer process, an industrial refining method. The Bayer process mixes crushed bauxite with the strong base sodium hydroxide under high temperatures to eventually separate aluminum containing chemical compounds from molecules within bauxite.[16]

Jamaica has been home to a number of successful bauxite mining operations since the 1950s, with the nation becoming one of the world's leading suppliers of aluminum, at least until a decline in the 1980s occurred due to a worldwide surplus. Aluminum compounds are the major constituent of the mineral bauxite, representing over 50 percent of the content of a typical sample. While aluminum is by no means a rare metal—in fact, it is the third-most abundant element in the planet's crust and the most abundant metal by far—its light weight, tremendous strength, and low cost are leveraged for use in everything from soft drink cans to airplane parts and candy wrappers.

In addition to aluminum, red mud features compounds containing titanium and silicon, along with trace quantities of scandium and rare earth metals.[17] Red mud also contains substantial quantities of iron, with these oxidized iron atoms giving the substance its namesake red hue. Despite the rich color, there is not enough iron content present in the mud to make further refining to acquire solid iron worthwhile. Since the creation of red mud is an economic dead end, the liquid is often left in human-made holding ponds to sit for years, changing from a slurry to a solid dirt form and back again through cold spells and rainy seasons. These open-air holding ponds render the land on which they sit unfit for housing or agriculture due to the extremely high pH (acidic nature) of the bauxite residue, injecting a chicken pox–like array of open wounds on the landscape of the Caribbean island. Treating the red mud with ocean water as the waste is created decreases the high pH of the mud, as the sodium and calcium salts in the seawater are able to react with metal compounds floating in the slurry.[18] This simple treatment method makes red mud far easier to dispose of, but unfortunately the treatment process was not well known in the past, leaving decades' worth of waste open to the air and environment in large holding ponds.

Red mud holding ponds are often placed in regions with plenty of open land and an accompanying low-population density with little or no danger to human life. When the boundaries of holding ponds are breached—rare as a disaster such as this may be—the resulting torrent causes immense physical and environmental destruction. A small breach in a series of weakened reservoir walls led to a 2010 accident at the Ajkai Timföldgyár refining plant in Hungary. The red mud quickly escaped the holding pond, with the resulting flood swallowing the nearby village of Kolontár with six-foot waves, killing ten people and eliminating all aquatic life in connecting waterways.[19]

The environmental impact to the island of Jamaica to recover the rare earth–laden mud would be minimal—red mud is already a blight on the landscape, and any red mud remaining after the collection of rare earths could be pumped back into the holding pond from which it was taken. Granted, this is not the most environmentally conscious route, but the further refining of red mud could create a "net good" by making use of the stagnant waste to quench some of the world's thirst for rare metals and possibly provide Jamaica with a revenue stream. Attempts at commercial purification of red mud are becoming a reality, with a joint-operation between the government of Jamaica and the Nippon Light Metal Company of Japan paving the way for a small red mud processing plant that opened in late 2013.[20] Nippon Light Metal is covering the cost of all associated buildings as well as all operating expenses, suggesting that their hopes are high for this processing center's ability to retrieve rare earths from the waste. Reevaluating the content of red mud in areas with a history of bauxite mining could lead to Jamaica, Australia, Vietnam, and West Africa's Republic of Guinea gaining a foothold through increased metal export revenues and possibly becoming much-needed alternative sources for rare earth metals.

DREDGING THE OCEAN FLOOR

Turning large holding ponds of red mud into profitable extraction sites for rare metals is an innovative approach, but Jamaica is not the only nation taking alternative paths to acquire a steady source of much-needed metals. Japan, arguably the most scientifically advanced nation on the planet, could resort to scouring the ocean floor for rare metals in the near future.

To fully understand Japan's desire to acquire its own long-term source of rare earth metals, one must consider centuries of conflict between the country and the world's rare earth magnate, China. The relationship ups and downs between China and Japan mirror a strife-plagued couple that has stayed "married for the kids" over the better part of two thousand years. China is often viewed as the historical figurehead of Asian civilization, leaning on a bevy of smaller nations—Korea, Vietnam, Indonesia, and sometimes Japan—to exploit its influence in the region.

Japan and China's early interactions have gone from amicable juxtaposition to ferocious modern struggles over land ownership and export regulations. The first recorded interaction between the two countries is a 57 CE gift exchange that saw the Emperor of China's Han Dynasty bestowing a three-and-a-half-ounce gold seal on the king of Na, the ruler of an area now incorporated into Japan. This was not a mythical gift exchange; the serpent-emblazoned seal was discovered in the eighteenth century and placed on display at the Fukuoka City Museum.

The two countries behaved cordially for centuries, with a sea battle in 663 CE briefly interrupting this era of peace. Rogue Japanese pirates preyed on Chinese villages from time to time over the next thousand years, but these appear to be the actions of individuals, as the pirate attacks also plagued the coasts of Japan and Korea.

While relations between Japan and China were cordial, China firmly held the upper hand. The remaining powers in the region constituting modern-day Vietnam, Korea, and Indonesia continually deferred to China and followed the country's lead.

The often hostile modern relations between China and Japan are rooted in military conflicts of the nineteenth and twentieth centuries. The First Sino-Japanese War pitted the ruling families of China and Japan against one another for control of an unstable Korea. Japan's intentions for Korea were not completely benign. Japan coveted Korea's coal and iron supplies, materials much needed if Japan was to leapfrog into the Industrial Era. The battles began in August of 1894, with the war lasting just shy of nine months and ending with China's ruling Qing Dynasty falling on poor military footing and eventually suing for peace. The war marked the beginning of Japanese military dominance in the region, as the much smaller country—both in land mass and population—soundly defeated the Qing Dynasty (China's population was nearly four hundred million by the end of the nineteenth century, more than ten times that of Japan).[21]

Japan benefited from a greatly weakened opponent. China and the Qing Dynasty struggled against one another during the First and Second Opium Wars, making for fourteen years of civil war in southern China. China also fought a number of battles with France over custody of what is now northern Vietnam only a year before the dawn of the First Sino-Japanese War.

At least a part of Japan's rapid evolution into a military and scientific power during the nineteenth century is due to a policy that welcomed Western culture and, specifically, influence on the part of the United States after the reign of shoguns who favored seclusion. China continued to fall away in following decades, with historians deeming the last half of the nineteenth century until the end of World War II a "Century of Humiliation" for China.[22]

As China's influence receded, Japan grew in strength, with this smaller country protecting the broad interests of Korea, Vietnam, and China. Japan entered into war with Russia in 1904, a war wherein Japan once again defeated a much larger opponent as it succeeded in preventing Russian influences from taking physical possession of a port in the region.

Bolstered by recent successes, Japan flexed its military and political might in the region through 1915 and issued the Twenty-One Demands, a series of declarations directed toward the newly formed Republic of China aimed at protecting Japanese interests in China proper and the region as a whole. The final blow that set the stage for their strained modern-day relations came in

1937 with the beginning of the Second Sino-Japanese War. This second large-scale conflict between Japan and China is often lost on those with a Westernized view of history—a view that focuses on Europe and northern Asia in discussions of the 1930s and 1940s due to World War II.

The Second Sino-Japanese War is punctuated by the "Rape of Nanking": six weeks of terror in 1937 that followed the Japanese capture of the capital city of the Republic of China. Chinese civilians were not outside the purview of the Imperial Japanese Army, with an estimated three hundred thousand Chinese civilians killed while Japan held Nanking. Japanese attacks on Chinese civilians were unfortunately routine due to the number of conflicts fought on Chinese soil, with the numbers bolstered due to air raids intentionally targeting Chinese cities in the hope of crushing China's resolve. China received aid from Germany, Russia, and the United States at various times during the war. Eventually the ties between nations led the Second Sino-Japanese War to bleed into World War II, particularly after the United States officially joined the Allied cause after Japan's bombing of Pearl Harbor.

The wounds of this World War II–era conflict run deep, with the scars left putrefying thanks to tales of horrifying and lethal scientific experiments carried out on Chinese nationals by Japan's infamous Unit 731. These experiments included forced insemination of Chinese women, intentional syphilis transmission, and tests to observe the impact of extreme temperature ranges on the human body. Unit 731 is also accused of directly attacking the Chinese people by preparing nonexplosive ceramic "bombs" that contained food staples contaminated with fleas that carried the bubonic plague.[23]

Japan and China are now in the midst of a cold war with a small set of islands located between the two nations at stake. Japan carried out a set of frequent "Island Defense" military exercises in the vicinity of the disputed Senkaku/Diaoyu Islands, which ignited tempers on a near-monthly basis in the early 2010s (the islands are known as the Senkaku to most of the world but are referred to as the Diaoyu in China). In the eyes of the Japanese government, the Senkaku Islands are without a doubt theirs to possess—Japan gained control of these islands in the final days of the nineteenth century as a portion of their spoils from the First Sino-Japanese War, with control passing to the United States after World War II until their return to Japan in 1972. The modern-day battle for the Senkaku Islands began during the 1972 transfer of governance, when China made a claim to their former territory.[24]

The Senkaku Islands themselves are unremarkable. The chain of five small islands and three uninhabitable rocks covers less than two thousand acres with not a single human living on the islands. The true value of these unimpressive islands is not in the islands themselves but in the oil and natural gas reserves that are thought to lie just under the ocean floor surrounding them in the South China Sea. A 1969 United Nations geological expedition pointed out the riches under the Senkaku Islands, with modern projections placing the oil supply at north of two hundred billion barrels, enough to supply China's growing petroleum needs for four decades.[25] The disputes are unfortunate for allies of Japan, who, much like in China's threat of embargo on rare earth metals in 2010, are also embargo targets over this dispute.

A full-scale military conflict in the coming decades between Japan and China is unlikely. Although Japan defeated China in the First Sino-Japanese War and held their own with China in the Second Sino-Japanese War until the latter became part of World War II, China's modernization over the last eight decades along with its massive population gives the country a distinct advantage. The Japanese Self-Defense Force can claim but a single member for every twelve members of the People's Liberation Army. If prolonged military conflict did take place, Japan's numbers and might would swell through likely contributions from the United States, with traditional allies like Australia and the United Kingdom also coming to its aid.

Charting China's allies is more difficult in the current political climate. Russian-Chinese relations run hot and cold, with both sides going toe-to-toe on economic battlefields for supremacy in the region. China may be able to rely on North Korea due to economic ties and shared borders, but China has voiced displeasure in recent years concerning North Korea's desire to continue nuclear weapons testing.[26]

The results of the Second Sino-Japanese War combined with the increase of Western influence in Japan in the aftermath of World War II led to the current frigid and suspicious relationship between China and Japan, a relationship that could very easily lead Japan to mine the bottom of the ocean around these islands in an effort to reduce its dependence on rare metal imports from China.

Polymetallic nodules—baseball-sized deposits of minerals and metals—line stretches of the ocean floor. These deposits form in a manner similar to pearls, with the layers of a polymetallic nodule slowly added over time, while the cores are created out of bits of bone or tooth from recently deceased sea life.[27] The layers are primarily composed of manganese, but copper and cobalt are also present, along with small amounts of eleven of the seventeen rare earth elements.[28]

While pearls can be grown and harvested in a few short years, polymetallic nodules grow a mere half an inch in diameter over the course of a *million years*—not exactly the timetable we see with renewable resources. Once the last manganese nodule is harvested and refined, that will be the end of underwater rare metal mining.[29]

These nodules dot the bottom of large segments of the Pacific and Indian Oceans, sitting just below the sand.[30] Selectively picking up these fist-sized pieces of rare metal–containing rock without destroying the seafloor would be an environmentally friendly strategy, but a surgical extraction of this sort is not possible due to the sheer number of nodules needed for a successful mining operation.

While polymetallic nodules form layer by minuscule layer at a snail's pace, this is far from the biggest logistical problem plaguing large-scale methods to acquire these ancient deposits. Polymetallic nodules are most prevalent at depths of two to three miles below the ocean surface, depths that require bravery and ingenuity to scour.

Using manpower and money to dredge the bottom of the seas for metal-containing nodules came to the forefront for a tiny window of time during the 1970s. Eccentric billionaire and captain of industry Howard Hughes kick-started the trend by announcing plans to build a three-hundred-and-fifty-million-dollar deep-sea drilling ship in an effort to make a dent in the world's surplus of manganese nodules. The *Hughes Glomar Explorer* left port in 1974 for an expanse in the Pacific Ocean thousands of miles north of Hawaii. It's unfortunate for modern underwater mining that the entire venture was a lie.

The Central Intelligence Agency was desperate for a cover story (and the proper vessel) to use for its upcoming Project Azorian, an investigation and retrieval mission centering on a Soviet Union K-129 nuclear submarine that sank in 1968. The Soviets knew the general location of the sunken vessel and made several attempts to retrieve the nuclear missiles onboard as well as the classified information and the bodies of crew members, but with no success. A forensic recovery project that large would require a new generation of equipment the US government could covertly manufacture and assemble, thanks to a willing Howard Hughes.[31]

The massive hull of the *Glomar Explorer* hid a retractable claw designed to retrieve the K-129 submarine, just one of the many tricks the CIA concealed up its collective sleeve. The CIA nearly succeeded in recovering the entire hull of the Soviet K-129, only to experience half the submarine breaking off and drifting back to the ocean floor before reaching the surface. Although a large portion of the sub was lost, a set of Soviet nuclear missiles were recovered during the extraction, providing the United States with a wealth of insight about the Soviet Union's military capabilities and salvaging the efforts of the entire operation.

The head of the CIA during Project Azorian would go on to become one of the most prominent political figures of the late twentieth century: George Herbert Walker Bush. The CIA fought hard to keep Project Azorian a secret in order to prevent an international incident in the midst of the Cold War, going so far as to suppress a *New York Times* piece detailing the true purpose of Azorian.[32]

If the Soviet Union became prematurely aware of a US effort to retrieve the submarine wreckage, any hopes of peaceful talks between the countries would have evaporated. In a gesture of goodwill, the United States gave several of the deceased crew members a proper burial at sea, sending the Soviets documentary footage of the burial well after tensions between the two countries had relaxed.[33]

If you are concerned about the continuing story of the *Hughes Glomar Explorer*, you can relax—this massive piece of secret US history was neither mothballed nor ripped apart for scrap metal sold and scattered to the ends of the earth. In the years after Project Azorian, Shell, Lockheed Martin, and the now-defunct Standard Oil Company leased the *Glomar Explorer* from the US Navy and made attempts at privatized ocean mining.[34]

If and when nodule mining becomes a reality, the process will build upon the existing foundation put in place through the underwater mining of diamonds. The De Beers Corporation currently operates five full-time vessels for this purpose, with all five dedicated to sifting through shallow sediment beds off the coast of the African country of Namibia.[35] The German-based company found underwater operations far more efficient than above-ground mining efforts, as a fifty-man crew armed with state-of-the-art technology can match the output of three thousand traditional mine workers.[36]

Two methods used for underwater diamond mining are directly

applicable to retrieving manganese nodules from the ocean floor. Drilling directly into the seabed is a possible retrieval option, with this avenue penetrating deep below the floor to bring up broken-up rock, sediment, and nodules through alien-looking, mile-long tubes. Once the debris is brought to the hull of a mining ship, chemical and physical processes are used sift to through the cargo, with any undesired rock and sediment returned to the bottom of the ocean floor. The second method shuns the use of drilling and instead uses a combination of conveyor belts and hydraulic tubes to cover larger areas than are accessible by drilling.

In both situations effort must be made to return unused sediment and rocks to the site from which they were displaced. This environmentally conscious factor will weigh heavily on any underwater mining operation since it is impossible to return the disturbed and sifted sediment and rocks to its original orientation. Adversaries of underwater mining will have several clips full of ammunition at their disposal. It will be easy to make a case blaming underwater mining for any and every adverse biological event stemming from the displacement of soil and aquatic life during the course of mining. Pushing the burden for remediation onto mining corporations could be a difficult task because the companies will, in most cases, have moved on from the site long before any damage becomes apparent while directly pinning blame will be a difficult task as governments and corporations could continually cite that mining operations took place within a short window compared to the several-billion-year lifespan of the ocean floor.[37]

Despite the effort and controversy, deep-sea mining will be worthwhile for many nations. Japan is not alone in its interest: India's National Institute of Oceanography is actively studying polymetallic nodules in order to acquire its own strategic reserve of rare metals from the surrounding Indian Ocean.[38]

While India's push is driven by economic indicators, Japan's desire is fueled by fear: that its technology-entrenched society will be stunted by a hard embargo on the part of China. Taking nineteenth- and twentieth-century conflicts into account, such a maneuver would not be out of the question since it would rightly fit the current (and likely future) political climate between the two countries. A long-term embargo of crucial metal exports from China, the world's leading provider, would send Japan across the globe at great expense to accumulate a working supply. China experimented with a short-term embargo of rare earth metals in 2010, preventing shipment of the metals en route to the island nation in response to the Japanese Coast Guard detaining the captain of a Chinese fishing vessel found operating in the vicinity of the Senkaku Islands.[39] The bureaucratic process, as one can expect, did not move quickly; a World Trade Organization panel handed down a ruling an astonishing four years after the incident. Despite the delay, the World Trade Organization's dispute panel ruled against China's action.[40] During this span of time the financial markets tied to these metals fluctuated wildly in the midst of a questionable future, with fears that China might one day cut off the rest of the world from its massive supply, leading Canadian- and United States–based corporations to reopen investigations of deposits within their borders.

Underwater mining will face significant environmental resistance should the activity become commonplace. The impact of a slightly similar process, bottom trawling, is well known, and the negative effects associated with this fishing practice will no doubt be used as ammunition against even the most benign practitioners of polymetallic nodule mining. Bottom trawling uses a weighted net to dredge the bottom of the seafloor for fish and crustaceans, with a large portion of the world's dietary portion of halibut, flounder, and cod, along with shrimp and squid, coming by this method. Like underwater mining methods that would vacuum nodules from the seafloor, bottom trawling disturbs the ocean floor, kicking up sediment and clouding the area with trails visible by satellite.

Gleaning the ocean floor, sifting through toxic mud, and breaking through the icy surface of Greenland and Antarctica are radical alternatives to traditional mining. Despite their unusual nature, each option could become a reality should our need overwhelm the current reserves of rare and scarce metals or an interruption occur in the current supply chain.

These alternatives, however, are tame compared to unorthodox approaches future generations may use to combat a scarcity of necessary metals—a combination of need and technology could send humankind to other planets or to the stars to quench our thirst for scarce metals.

CHAPTER 16

GOING THE DISTANCE

If the desires of humanity exhaust Planet Earth's entire supply of rare and necessary metals, what does our future hold? Will we be forced into an endless circle of recycling and rationing of technology? Maybe not. If spacefaring technology catches up with our global need, we will find all the metals we require within the moons and asteroids of our solar system. One question, however, will persist: who lays claim to space?

WHO OWNS SPACE?

The Apollo lunar missions left six flags on the cratered surface of Earth's moon, but are any of the flags still standing? A study of images taken by the *Lunar Reconnaissance Orbiter* in 2012 showed five of the folding Lunar Flag Assemblies standing decades later. Why just five? Buzz Aldrin saw the flag left by *Apollo 11* fall due to rumbles felt after the return rocket liftoff. NASA scientists questioned how the flags would endure the intervening years, positing that they would disintegrate quickly under the forty-two-year barrage of radiation, tiny meteorites, and fifteen-hundred-plus instances of five-hundred-degree temperature swings the flags have endured due to the broad differences in lunar surface temperatures between night and day.[1] While five flags still stand on its surface, do they mean anything in regard to possession of the moon? The bold image of armor-clad members of a nation-state planting a flag in the soil certainly makes a statement. The act is a centuries-old sign of ownership, and one conjured up over and over again in elementary school when learning about conquistadors' claims of lands in the New World for their home country. If humanity reaches the point where we begin to look outside of our atmosphere for resources, could spacefaring nations revisit the travels of decades ago to make claims in their favor?

Mining extraterrestrial objects, whether on the moon or an asteroid, brings about a number of political questions and a few legal ones, too, for attorneys dabbling in extraterrestrial real estate. Nations and private citizens have claimed ownership of the moon, with the United Nations putting forth the poorly received *Agreement Governing the Activities of States on the Moon and Other Celestial Bodies* in 1979. The treaty states that all nations enjoy a joint-claim to bodies outside the boundaries of our planet, but as of today, only nineteen countries have supported the treaty in the nearly four decades since the United Nations opened it for signing. The three leading spacefaring countries on the planet—China, the Russian Federation, and the United States of America—are noticeably absent.[2]

Why would countries care about a physical stake in the moon, a satellite that appears to be a barren mass of gray rock, or any other celestial body? Looking to the stars for atomic wealth is not an unreasonable notion since the fiery centers of a billion suns were the setting for the creation of each element from carbon to uranium. Only a handful of elements were present in the moments following the big bang—hydrogen, helium, lithium, and a small amount of beryllium existed. Creating stable nuclei at the starting line of the universe posed a problem with the available resources, but, as we now know, that the situation changed. (If not, you would not be sitting here as a carbon-based life-form and likely would not be sitting at all. Maybe existence would have taken on a helium-based gaseous state with intelligent life-forms floating gently through the universe.) Stability issues led to the creation of only extremely light elements during the big bang. As the universe expanded and star formation began, the elements from carbon to iron on the periodic table were ushered into existence through collisions made possible due to the intense heat, gravity, and sheer density of atoms present at the core of stars.

Why did this act of astronomic creation stop with metallic iron? The creation of iron atoms inside a star marks a fateful moment, with continued build-up of iron atoms leading to the death of the star. Stars burn through the release of energy from the ongoing fusion of atoms. Iron is capable of fusion, but it does not produce any energy. Iron actually needs an input of energy to undergo fusion. This energy is pulled from energy-producing fusion reactions in the core of the star. As more iron atoms are created through fusion and then undergo fusion themselves, more energy is needed from the surroundings, eventually cooling the core temperature of the star and initiating a death spiral.

We, however, are the benefactors of this death spiral—as the first stars began forming iron and started down the path to going supernova, a different block of elements were created for the first time. The contraction of stars led to the creation of most elements on the periodic table. These elements, born out of the death of stars, are creations prized above all others after the basic building blocks of life like nitrogen, oxygen, hydrogen, phosphate, and carbon are present and accounted for.

SOMEONE ALREADY OWNS THE MOON?

Ownership of what lies outside of Planet Earth is a touchy issue, with spacefaring nations, thus far, failing to extend the spirit of the existing United Nations Convention on the Law of the Sea to the space outside of our planet's atmosphere (although certain rights are extended to the areas where geosynchronous satellites reside over a given country). If extended to space, the existing tenants of the UN convention would forbid all governments from claiming ownership of land on extraterrestrial bodies. While national claims to the moon are in limbo, private citizens routinely make formal claims to the moon and other celestial objects, even going so far as to sell real estate on the moon.

Claiming possession of celestial bodies is not a modern phenomenon. Ancestors of eighteenth-century German healer Aul Juergens claimed in the late 1990s that King Frederick II of Prussia, better known as Frederick the Great and infamously remembered due to Adolf Hitler's near-idolatry of his reign, set aside the moon and orchestrated a familial succession of ownership in return for Juergens's years of loyal service to the crown.[3] At roughly the same time, a trio of Yemini men made worldwide news for their attempts to sell parcels of Martian real estate for the low, low sum of two dollars per square meter. These men entered the space realty business after claiming that their ancestors, members of the Himyarite and Sabaean Kingdoms in Yemen, rightfully claimed ownership of Mars thousands of years ago. The low price point of Martian land paled in comparison to current real estate prices in their homeland—a corresponding parcel of land in the downtown area of Yemen's capital, Sana'a, would cost five hundred dollars per square meter. Not satisfied with their current level of audacity, the Yemeni three followed up their announcement and sales plan by suing the North American Space Agency for trespassing after the unmanned *Pathfinder* and *Sojourner* crafts landed on Mars in July of 1997.[4]

Depending on one's plans for future trips to Mars and one's faith in Yemen's legal system and a lawyer's ability to enforce private property claims from one hundred million miles away, purchasing a set of the Martian parcels might prove to be a shrewd real estate investment. There are far worse ways to spend two bucks.

European and Middle Eastern descendants looking back from the limbs of family trees are not the only ones laying claim to extraterrestrial bodies and attempting to make a financial return on their perceived possession. Moon Estates, a company operated by a gentleman in his late sixties from California, is more than willing to sell a tract of lunar property to anyone. Moon Estates' Dennis Hope claims to have sold six hundred million acres of land on the surface of the moon since 1980.[5] Moon Estates includes deed certificates to go along with these quaint gifts of lunar land that seem perfect for a cheesy pass at a science-loving coed. After laying claim to not just the moon but all planets in the solar system through a letter to the United Nations, Hope began the process of gridding the moon and doling out one-acre lots to anyone willing to send him the paltry sum of twenty dollars plus taxes and fees.

A first-year law student could skillfully unwrap the tales and question whether Frederick the Great or ancient inhabitants of modern Yemen held the ownership over the moon or Mars, let alone attempts to transfer ownership in perpetuity. Even if an individual could make a case for ownership and transmission, state and national claims would supersede their own. While it makes for interesting fodder, if and when the time comes to parcel out land on Mars and the moon in the future, these descendants will experience a significant amount of push-back if and when they attempt to enforce ownership.

While selling land on the moon or Mars borders on taboo, selling items currently on the earth's primary satellite is within reason for major auction houses contracted by prominent governments. In 1993, Sotheby's auctioned off a Soviet lunar rover, the *Lunokhod 2.* One important problem prevented delivery of video game magnate Richard Garriott's prize—the *Lunokhod 2* is marooned on the surface of the moon.

Garriott's purchase—albeit made possible by the breakdown of a former superpower and a circuitous path of events—marks the first time a private citizen held legal claim to an object in space. Richard's father, Owen, is a decorated scientist and astronaut who flew several missions with NASA, including a sixty-day stint during the third Skylab mission, which gave the elder

Garriott and the Skylab 3 crew the record for the longest space flight duration. His son intended to follow in his father's footsteps, but Richard's interest took him on a different path, as he became one of the first celebrity video game developers. Garriott is responsible for the Ultima series of role-playing games and is beloved by fans for his willful immersion into fandom, where he is fondly known for choosing to cast himself as the in-game ruler of his world, Lord British, in Ultima and Ultima Online.

Armed with a bottomless checkbook and a lifelong love of science and the final frontier, Garriott purchased the *Lunokhod 2* for $68,500 in 1993 from Sotheby's New York. Lavochkin, a state-owned Russian aerospace company, signed the rover to Sotheby's in an auction devoted to selling memorabilia from the Russian space program to North American bidders just over twenty years ago. In addition to the 239,000 miles separating Garriott from the *Lunokhod 2*, Lavochkin did not know the exact location of the rover, since the Soviet Union lost communications with the craft in 1973.[6]

The *Lunokhod 2* roamed around the surface of the moon leaving tracks in its wake for over three months and performed several observation missions. A combination of lunar dust and an exposed radiator led to an overheating problem that caused *Lunokhod 2* to cease functioning. Although neither Garriott, Sotheby's, nor Russia's Lavochkin knew the position of *Lunokhod 2* at the time of the auction, the increased attention paid to the rover led to its "rediscovery" in 2012, when a University of Western Ontario professor found *Lunokhod 2* as he sifted through newly released photos taken by NASA's *Lunar Reconnaissance Orbiter*.[7]

The *Lunokhod 2* is far from the only spacecraft sitting on the moon; close to seventy-five other craft in various states of assembly reside on the surface, abandoned for decades. What is unique about the *Lunokhod 2*? The purchase made Garriott the sole publicly known private owner of an item outside of Earth's orbit.

The *Lunokhod 2* is not the only craft on the moon that Garriott owns. He is also owner of the accompanying landing craft from the 1973 Soviet mission, *Luna 21*, which orbited the moon and landed on its surface prior to successfully deploying the *Lunokhod 2* rover for an exploratory mission. *Luna 21* remains on the moon at its original landing position of 25.85° north latitude, 30.45° east longitude within the lava-worn Le Monnier crater.[8]

After video game cornerstone Electronic Arts purchased Garriott's development studio, Garriott bought a ride on a 2008 Russian mission to the International Space Station. The bill for this trip? Thirty million dollars, or four hundred and thirty-eight times the price of the *Lunokhod 2*.[9] The trip not only marked the culmination of a lifelong dream for Garriott, but it propelled him into the history books as the first second-generation astronaut in human history.

Why an agency of the former Soviet Union was legally able to auction off long-abandoned items on the moon's surface is a compelling question itself, but the sale of *Lunokhod 2* makes for an interesting personal property case. Garriott *legally* owns property on the moon. The rover and landing craft are not real estate, but do the purchase of these spacecraft constitute the beginning of a "claim" to the land in their vicinity? Flags placed on coasts of the New World denoted a country's claim, and in a political atmosphere wherein treaties prevent individual governments from acquiring lunar territory, future courts may decide that ownership of *Luna 21* and *Lunokhod 2* gives Garriott (and/or his heirs) rights to a piece of *la luna*—a nice low-risk, high-reward investment for a billionaire space enthusiast.

Sotheby's auction house continues to sell a number of Russian space artifacts, with a hulking spherical *Vostok* capsule similar to the one used weeks later by Soviet cosmonaut Yuri Gagarin as he became the first human in space, selling for $2.9 million in 2011.[10]

The moon is large, but nowhere near the size of Earth—its surface area is smaller than the entire continent of Asia. It features a thin covering of metal-containing dust and soil along its rocky surface, a mixture given the name regolith—Greek for "rock blanket"—by geologists. The Apollo missions brought a combined 842 pounds of rocks back from the moon, contributing to the knowledge base by giving us a very good idea of the satellite's composition.[11]

I doubt that mining missions rendezvousing on the moon will center on the acquisition of rare metals since the regolith-cover does not appear to hide large quantities of desirable metals. This does not mean the celestial body will lack a role in potential mining operations in our solar system—reality will be just the opposite, with the moon providing an important base of operations for these voyages. NASA studied the composition of regolith for the past five decades in the hope of using the covering as an energy resource. Regolith is rich in helium-3, an isotope of helium featuring one less neutron than helium present in abundance on Earth, which is a potential clean energy source for use on Earth or by lunar bases. Helium-3 readily enters into two safe and efficient fusion reactions without the formation of radioactive waste.

The sun continually releases helium-3 and sends the isotope whirling around the planets on solar winds. Unfortunately for us, very little helium-3 is found naturally on Earth because the magnetic field of our planet pushes away these useful helium atoms. This is not a complete loss since the magnetic field deflects cosmic rays that would otherwise render Earth inhabitable. The moon exhibits a very weak magnetic field, allowing helium-3 to build up within its rocky soil for billions of years.

Potential commercial and government efforts to reach the moon or the many asteroids in its vicinity and beyond could leverage residual helium-3 within lunar regolith. Wise caretaking of regolith could provide a commercial product for transport to Earth (making reality of the plot of the eerie 2009 science fiction drama *Moon*) or an energy source useful to life-support facilities and lunar bases acting as a way-station on large-scale mining operations. Helium-3 is not the only useful resource awaiting humankind on the surface of the moon. Experiments show that the regolith itself is capable of use as a building material. Making use of existing materials on the moon for construction would eliminate a considerable portion of long-term operation costs, and to explore this, NASA loaned out several small samples of lunar material returned from the Apollo missions to academic and corporate labs. These labs successfully created a stable cube of regolith and rods approximating structural beams, taking the use of regolith as a building material from the realm of science fiction to science fact.[12]

TURNING TO ASTEROIDS

At first glance, asteroids may appear to be indeterminate boulders floating through space. The average person is aware of their existence in faraway stretches of space, but the only time an asteroid enters his day-to-day life is as part of the plot device in disaster movies (*Deep Impact* and *Armageddon* are two of my guilty pleasures) and the rare thirty-second science segment on the nightly news reporting the discovery of an asteroid that could, but likely will not, strike the planet decades in the future. When you hear the latter remember that asteroid collisions are not an entirely negative event—without the fateful asteroid strike that left a one-hundred-and-eighty-kilometer-wide crater in the Gulf of Mexico sixty-five million years ago, human beings could be competing with herds of triceratops, the occasional velociraptor, and their evolutionary descendants for survival.

Asteroids are far more complex than the generic boulder. Astronomers divide asteroids into groups depending on their shape, proximity to Planet Earth, and their albedo (the ability for the surface to reflect light). Using this system, researchers have identified over six hundred thousand asteroids (or, if you are a disaster movie aficionado, bullets in the chamber of the Universe's favorite revolver, ready to kill us all at a moment's notice in a cinematically pleasing fashion) large enough to receive designations from the International Astronomical Union that pass near the vicinity of Earth.

The exterior of asteroids are a very a harsh environment, different from the surface of Earth due to the lack of any atmosphere. Asteroids lack the mass and size necessary to keep the tiny molecules needed to sustain an atmosphere within their gravitational pull, so their surface is bombarded with every iota of radiation that our atmosphere thankfully protects us from.

We know a small amount about the composition of asteroids from direct observations. Recovered meteorites—asteroids that make it through the planet's atmosphere—are largely made of iron and nickel. Recent analysis of well-preserved four-billion-year-old extraterrestrial rock samples from ice-covered portions of Greenland show differences in tungsten and neodymium concentration when compared to terrestrial rocks exposed on the surface.[13] This study indirectly gives validity to theories that meteorites contained minute quantities of gold, tungsten, and other scarce metals that are known to be "iron-loving" metals, with these metals brought by the iron/nickel meteorites supplementing the planet's native supply through countless collisions over millions of years.[14]

With the size of an asteroid pinned down and a reasonable idea of the composition of the surface in hand, it is natural that financial prospects come into play. Estimates of the cash value, taking gross and net profits into account, of over one thousand well-characterized asteroids exist, allowing armchair space venture capitalists to make guesses as to which asteroid humankind will mine first.

The profitability of individual known asteroids in our solar system is debated and ranked, with those named 4034 Vishnu and 2000 BM19 estimated to contain more than twenty trillion dollars of material each and likely mining targets due to their near-Earth proximity, small size (both are less than a kilometer in diameter), and projected profitability.[15]

The first unmanned asteroid visit has already taken place, but the mission has been overlooked by many. Japan, a country desperately in need of rare earth metals to provide a stable supply for consumer use and manufacturing, developed and launched the *Hayabusa* in 2003 to visit a preselected asteroid and return home with a sample in tow. The *Hayabusa* made use of a novel propulsion technology that will be essential to asteroid mining: ion drive engines. While real-life application of ion drive engines has occurred only in the past two decades, *Star Wars* mastermind George Lucas has used the theory behind ion drives to explain how TIE Fighters and Star Destroyers moved through space in the movie saga (the TIE Fighter is named for the technology: *T*win *I*on *E*ngine). The *Hayabusa* traveled to the asteroid and back to Earth by ionizing molecules of xenon and ejecting the ions in the opposite direction of the craft.[16] This is the hybrid automobile approach to spaceflight, with only a small amount of xenon fuel necessary to move the *Hayabusa* along the desired contact route. The downside to making use of current ion drive technology is speed. Quick changes in direction are nearly impossible with ion drives, so the engines were supplemented by chemical propellants to bring the *Hayabusa* back on course in emergency situations.

The *Hayabusa* hovered near the asteroid 25143 Itokawa—a well-analyzed asteroid that regularly passes near Earth—and made two sample collection runs. The craft landed on 25143 Itokawa with the intent of shooting projectiles at the asteroid's surface to dislodge debris for collection and return to scientists back home.[17] Yes, the *Hayabusa* was to essentially shoot bullets at the asteroid, but the act was in the name of science, so all is forgiven.

Problems with the projectile launcher and sample collection apparatus prevented the return of a substantial sample, with the only matter from 25143 Itokawa returned for analysis stemming from particles unintentionally lodged inside of *Hayabusa*. A nice surprise for a mission once thought lost through bad luck.

Hayabusa project planners at the Japanese Aerospace Exploration Agency will make use of explosives in future asteroid rendezvous attempts. *Hayabusa II*, which should be en route to its destination by 2018, will carry a payload that includes over ten pounds of cyclotetramethylenetetranitramine, a military-grade explosive.[18] Why does a small unmanned spacecraft need enough explosives to level an office building? The explosives will allow *Hayabusa II* to collect fresh, untouched samples of an asteroid by creating a crater on its surface.

You may have a question buzzing around in the back of your head right now: if humans have never set foot on an asteroid and our only direct observations come from tiny fragments landing across the globe, how can a dollar value be applied to its mining resources? Astronomers make use of information gained from the absorption of light by a particular asteroid to determine its likely composition. While performing this task, astronomers use methods quite similar to the ones employed by nineteenth-century chemists and geologists looking for new metals. The positions of these asteroids are very stable because they follow paths and patterns scientists can study over long periods of time. The composition of an asteroid is determined in part by the technique of "telescopic reflectance spectroscopy," which uses light to gain a better idea of the elements inside of these celestial rocks in a manner not too far removed from the methods used by Carl Mosander to describe the "false" element didymium and the three legitimate elements he discovered. While the theory behind telescopic reflectance spectroscopy is complex, the process is relatively simple in practice. Astronomers bounce light off of an asteroid and measure the amount of light the surface reflects. Once enough reflections are collected for a single asteroid, the information is compared to a library of reflectivity values for samples with a known mineral and element composition found on Earth, along with data obtained from analyzing meteorites that survived the trip through the atmosphere. At the heart of this comparison is the rationale that if a library sample and an asteroid have the same reflectivity values, their composition will be the same, allowing astronomers to determine the composition for asteroids and planetoids (a vague classification given to identifiable celestial bodies found in regular orbits in our solar system) tens of millions of miles away from Earth.[19]

EARLY VOYAGES

So where in our solar system will we send the first generations of "astrominers"? This brave bunch will likely be looking forward to a long trip, one that takes them over four hundred million miles away and to the edge of the bountiful Mars-Jupiter asteroid belt. Targeting an asteroid in proximity to our planet, even if it is projected to be worth a fraction of those found in the asteroid belt, would be a prudent option. Granted, the quality of the metals and other resources garnered would be decreased, but mining efforts need a practical starting point. Even with a government partnership, overcoming the difficulties inherent in a long-distance space journey in addition to the unforeseen hurdles no doubt in store on the first handful of attempts before venturing out will be challenge enough to keep humankind busy for years before tackling more profitable asteroids and planetoids that reside deep in our solar system.

The most widely read organizing system for asteroids is the Tholen classification, with the overwhelming majority of asteroids in our solar system fitting into one of three categories: s-type, c-type, or m-type. Each category takes its designator from the primary characteristic of the group, for example, the s-type is dominated by asteroids with a "stony" appearance. These asteroids consist of silicon compounds, the surface of these asteroids is often smooth and featureless, similar in appearance to the surface of the moon. The asteroid named 25143 Itokawa, targeted by the Japanese probe *Hayabusa*, is an s-type asteroid.

C-type asteroids are predominantly carbon, akin to giant pieces of coal hurtling through space. C-type asteroids populate the interior of the Mars-Jupiter asteroid belt, far from the rays of the sun. Preliminary data leads astronomers to surmise that c-type asteroids will hold hidden stores of water in their interiors bound to minerals similar to how water is found bound to clay on Earth. If this hypothesis is true, space travelers in the coming centuries may rely on asteroids as a life-support device, rounding up asteroids like cattle in the Wild West and gently transporting them to nearby bases for their water in a scheme that turns c-type asteroids into the *Ferocactus wislizeni* of space. Humankind can live in space without dysprosium, but no one can live for more than a handful of days without imbibing in a simple molecule consisting of two hydrogen atoms paired to an atom of oxygen.[20]

M-type (the "M" designation stands for metallic) asteroids are hypothesized to contain large iron, nickel, and cobalt deposits and a variety of other metals. These asteroids reflect light very well—if one was the same distance away as the moon, the asteroid would appear significantly brighter in the night sky.[21] While s-type and c-type asteroids contain large amounts of silicon and carbon, that does not mean rare earth metals or other valuable metals are absent. Our planet's surface is dominated by carbon, silicon, and other mineral-forming elements, yet plenty of metal abounds within.

WRANGLING ASTEROIDS

When I think of asteroids, I immediately go back to scenes of TIE Fighters slamming into asteroids as the *Millennium Falcon* successfully evaded their pursuit in *The Empire Strikes Back*. With this mental image in mind, we come to an important question for humans en route to an asteroid mining installation—will traveling to (and through) the asteroid belt be dangerous?

The Mars-Jupiter asteroid belt is home to millions of asteroids, with these celestial bodies varying in size from a speck of dust to four hundred kilometers across (the dwarf planet Ceres is also within the asteroid belt and measures an astonishing 950 kilometers in diameter). Even with what appears to be massive numbers, space is so vast that the belt itself is not densely populated. Unmanned spacecraft have successfully passed through the belt (*Pioneer 10* in 1972, *Voyager 1* in 1978, and *Huygens* in 2000, to name a few) on the way to Jupiter, Saturn, and the edge of our solar system, with no incidents occurring to date. The asteroids are spread out over a large enough area that collisions are unlikely to occur. This is a good development for humans and their spacecraft since the first mining vessels deployed in the asteroid belt will likely be unmanned, controlled either from Earth or a base closer to the belt itself.

Asteroids are not covered by the legal notions to which our satellite, our generically named moon, is held. Pieces of space rock hurtling through space also hold an added benefit for would-be miners: asteroids lack an atmosphere, dramatically decreasing the amount of fuel-mining missions that need to return to Earth. While missions to the Moon have shown the satellite to hold a trove of rare metals waiting to be used on Earth, the amount of fuel necessary immediately eliminates the majority of profit obtained through mining activities. The cost of taking any material—fuel, water, food, anything—into space is expensive, and leaving the atmosphere of the moon, however thin it may be, provides an additional financial burden. This burden combined with the political and ownership issues that arise from lunar mining will likely drive corporations to the mining of asteroids: extraterrestrial bodies lacking an atmosphere and nationalistic restrictions.

Towing an asteroid into the vicinity of Earth's orbit sounds like an idea pulled from a low-budget science fiction movie, but such a practice could become commonplace in our lifetime if sufficient financial benefit exists. The lasso used to wrangle and haul in an asteroid would be made not of steel and carbon fiber but rather from a series of small ion-drive thrusters. Ion-drive thrusters are used as a means of propulsion on unmanned spacecraft (NASA's *Dawn* and *Deep Space 1* probes) and move the probes through space by ejecting charged atoms of hydrogen, bismuth, and xenon. An array of ion-drive thrusters would need to be placed carefully on the asteroid and powered in tandem to move and direct large asteroids along, but the possibility is within the practical realm, thanks to existing technology. A second possibility for corralling and directing asteroids is the yet-to-be constructed Modular Asteroid Deflection Mission Ejector Node (MADMEN), which would attach to an asteroid and break a segment apart slowly. The MADMEN apparatus then propels the asteroid through the vacuum of space by ejecting broken pieces in the opposite direction of the desired course heading. The goal of MADMEN as funded by the NASA Institute for Advanced Concepts is to redirect an asteroid set on a collision course with Earth, but the same design would be useful for directing asteroids to the general vicinity of their mining destination as long as a small amount of the asteroid itself was broken apart and ejected en route.[22]

Once a propulsion system is in place on the asteroid, nudging the hulky mass in just the right place might pose a problem since any error could destroy the web of communication satellites that crisscross the sky. What if the unwanted but fathomable happened and a nation or corporation in command lost control of an asteroid and it hurtled down toward Earth? The mere possibility of such an accident could tag orbital asteroid mining as a military maneuver in the guise of a corporate operation due to the possible political fallout and damage incurred in the aftermath of a catastrophic mission failure. If a mishap occurred that led to the loss of life or property on Earth, would the nation where the corporation is based be liable for the damage? Disasters would be compounded if an incident harms individuals of a rival nation-state, further blurring the lines between corporations and nations while allowing foreign governments to point fingers and make accusations in a disastrous incident that might simply be an accident occurring one thousand miles above Earth's surface.

Logic dictates that one of our planet's foremost governments would be the first party to send a full-fledged mission into space to mine asteroids due to the enormous costs involved and the complex infrastructure necessary to put a small crew weighing a total of one thousand pounds into space, let alone return with several tons of cargo. The current transition period at NASA—years after the final Space Shuttle flight and long before the materialization of its successor—places bureaucratic and technological barriers in front of planning an asteroid mining flight. The dearth of government-initiated space ventures in North America has led corporations and private citizens, the majority of whom earned their fortune by helming world-changing web startups in the early part of the century, to take a turn at space travel, and with this interest, a handful of companies are entranced by the wealth theoretically available through the pillaging of asteroids. Google cofounders Larry Page and Eric Schmidt have teamed with filmmaker and explorer James Cameron to form Planetary Resources, Incorporated. Two of Cameron's films hold top-five positions in tallies of the highest-grossing films to date, with 2009's *Avatar* featuring a plot that mimics the tales of Central African mining conflicts, with a planet and its indigenous population threatened by a corporation's pursuit of a valuable metal paradoxically dubbed "unobtanium." While Cameron's unobtanium serves only to move the plot forward and acts as the ideal film MacGuffin, there is plenty of reality in this private venture—Planetary Resources has a launch agreement in place with Richard Branson's Virgin Galactic and hopes to have the first stage of its ARKYD series of mining satellites in position by 2020.

Unmanned exploration will be the avenue pursued in the early decades of commercial asteroid mining. The cost of placing humans in space—along with the necessary food and water to support them—runs around ten thousand dollars per pound. The price quickly enters the tens of millions of dollars to put the average-weight male into orbit, with many times that in food transport costs to keep the crew fed and hydrated during a prolonged journey. Robots, on the other hand, can be manufactured to carry out similar tasks, and only need a power supply to replenish their batteries at the end of the day. Strategies to send humans on the first extraterrestrial mining treks could easily be squashed on financial grounds long before safety, separation anxiety, and interpersonal relations play a role in commercial mission plans.[23] If the world runs out of rare metal resources, asteroids look to be an environmentally friendly but cost-prohibitive alternative with a slew of scientific roadblocks along the way. Breaking down an asteroid would have little to no negative environmental impact on Planet Earth. All debris would need to be cleared to prevent a domino effect of space junk that might create additional, smaller asteroids. If you've seen the opening scenes of the Academy Award–nominated 2013 film *Gravity*, you understand how a small amount of space junk traveling at high speeds has enormous destructive potential, with each piece of debris fueling a deadly chain reaction.

Whatever government or corporation arrives first, we "normal" citizens of Planet Earth should hope an altruistic attitude accompanies their lunar claim. Due to the exorbitant cost and risk inherent in a mission to the Mars-Jupiter asteroid belt, those who lay claim to asteroids will likely not exhibit generosity. To expect generosity from a corporation risking its financial existence—prototyping and preparing for a trip of this magnitude would likely cripple a corporation if unsuccessful—is folly.

It is also possible we may find a currently unknown metal in one of these asteroids, the product of vastly different geological movements and climates foreign to Planet Earth. Banking on finding a yet-to-be discovered elemental metal or a combination of metal nonexistent on Earth to make a commercial product is a fool's errand, but the profits in store—if just from novelty value alone—will be a boon for the lucky corporation.

The first company to successfully break down and transport asteroid remnants back to Earth for use will not only become one of the most powerful private entities in history but will likely be the first private corporation to enter the hundred trillion dollar club. Such wealth along with dominance over vital supply lines could combine to make the corporation the East India Company of its day, bestowing unprecedented power to shape land, lives, and law at the mining sites and the space-ways leading to the rare and scarce metals that humanity covets.

EPILOGUE

The elemental metals we are able to use today—those metals existing in large enough quantities to isolate—have long since been identified, thanks to the herculean efforts of world-class scientists who devoted their lives to filling in gaps on the periodic table.

As the number of discovered elements swells, a harrowing trend of instability among newly synthesized elements becomes apparent. As meitnerium, copernicium, and their siblings are slotted into the periodic table, it becomes apparent there are no undiscovered "super metals" lying in wait for curious physicists and chemists to stumble upon as they slam smaller elements and atomic particles together at tremendous velocities in billion-dollar laboratories. The elemental metals available to us at the moment will likely be the only ones available in the future, with ingenuity solely guiding new uses and applications instead of the chance discovery of new and useful metals. Only when the well of readily available metals runs dry will we find out whether humanity will squander the technological advances of the twentieth and twenty-first centuries, turn to ecologically damning efforts and even war to obtain desired metals, or invest the necessary intellectual capital in order to forge a peaceful solution and maintain the status quo.

Admittedly, some of the methods discussed in *Rare* are long-shots, perhaps more appropriate for the plot of a summer blockbuster than the pragmatic worlds of economics, geography, chemistry, and physics. That said, without the far-fetched ideas capturing the minds of devoted researchers, the annals of the twentieth century would be quite different. The plausibility by which we discern the means nations and corporations use to acquire desperately needed metals in the present and into the second half of the twenty-first century and beyond depends on one's personal view of the future and faith in humankind.

Is your future shiny, filled with gleaming towers, new technology, with the ills of the poor and sick eradicated? In order for this vision to be realized, humanity needs unfettered access to a litany of prized metals and a cornucopia of other resources.

Maybe you are a pessimist. Your future is neither clean nor bright but instead is a grim, dark world where nations across the globe wage war against neighboring countries in the shameless pursuit of clean water, fuel, scrap metal, and food. Smog fills the air, the sun rarely penetrates the clouds, and children clothed in tattered rags scurry through dilapidated buildings looking for useful scraps to take home.

But perhaps you fall into a third category, one where the future is mixed bag of probabilities. A place where many of the problems plaguing our day-to-day lives are eliminated but still hold a plethora of new and unforeseen complications rushing in to fill the void.

Regardless of your personal stance when it comes to envisioning the future—Pollyana, pessimist, or pragmatist—acquiring and making wise use of rare and desirable metals will play a role in our current struggle to access and acquire petroleum. Whether the acquisition comes at the cost of the environment, human lives, or political alliances is the question, with the answer coming in the decades to follow.

ACKNOWLEDGMENTS

Thanks first to my wife for her love, friendship, dedication, and ongoing support—she often believes in me more than I do in myself. Thanks as well to my twin daughters. The pair spent many an hour on my arms and in front of a laptop during their first several weeks of life as I finished this book. I will cherish those moments until I have no more memories left.

My gratitude goes out to my mother for her sacrifices and steadfastness, as well as my family and friends for their many efforts in my life and ongoing encouragement. Thanks in particular to Tripp Reynolds for making me talk about the book early in the morning when I was not exactly ready to. You kept me on my toes. I would also like to thank Laurie Norton, Wayne Garrett, and the other educators I came in contact with that guided me, intentionally or not, in this direction.

I wish to extend my appreciation to Laura Wood at FinePrint Literary Management for her perseverance and the effort she expelled, as well as Steven L. Mitchell at Prometheus Books for taking a chance on yet another author early in his career. I would also like to thank Julia DeGraf, Catherine Roberts-Abel, Brian McMahon, Meghan Quinn, Jade Zora Scibilia, Nicole Sommer-Lecht, Melissa Raé Shofner, Mark Hall, Lisa Michalski, and the rest of the good people at Prometheus Books for their assistance in editing, designing, and promoting *Rare*. If you are still reading the book at this point, thanks to you well.

NOTES

CHAPTER 2. WHAT IS RARE?

1. David M. Darst, *Portfolio Investment Opportunities in Precious Metals* (Wiley, 2013).

2. Junius P. Rodriguez, ed., *The Historical Encyclopedia of World Slavery*, 2 vols. (Santa Barbara, CA: ABC-CLIO, 1997), vol. 2, p. 449.

3. *Rare Earth Elements—Critical Resources for High Technology* (US Geological Survey, Fact Sheet 087-02).

4. Chiranjib Kumar Gupta and Nagaiyar Krishnamurthy, *Extractive Metallurgy of Rare Earths* (New York: CRC Press, 2005), p. 154.

CHAPTER 3. PLAYING THE LONG GAME

1. Marc Humphries, *Rare Earth Elements: The Global Supply Chain* (Congressional Research Service, 7-5700, R41347, December 16, 2013), pp. 3–4.

2. Chiranjib Kumar Gupta and Nagaiyar Krishnamurthy, *Extractive Metallurgy of Rare Earths* (New York: CRC Press, 2005), p. 132.

3. Tracey Schelmetic, "Are Hybrid Vehicle Manufacturers Shifting Gears Away from Rare Earth Elements?" *Industry Market Trends*, December 11, 2012, http://news.thomasnet.com/IMT/2012/12/11/are-hybrid-vehicle-manufacturers-shifting-gears-away-from-rare-earth-elements/ (accessed October 17, 2013).

4. Saqib Rahim, "Shifting from Arab Oil to China's Neodymium?" *Environment & Energy Publishing*, January 29, 2010, http://www.eenews.net/stories/87039 (accessed January 9, 2014).

CHAPTER 4. INSIDE A SINGLE ROCK

1. Mary Elvira Weeks, "The Discovery of the Elements Part XVI: The Rare Earth Elements," *Journal of Chemical Education* 9, no. 10 (1932): 1751.

2. Peter B. Dean and Kirsti I. Dean, "Sir Johan Gadolin of Turku: The

Grandfather of Gadolinium," *Academic Radiology* 3, supplement 2 (August 1996): S165–9.

3. James L. Marshall and Virginia R. Marshall, "Rediscovery of the Elements: Yttrium and Johan Gadolin," *Hexagon* (Spring 2008): 9–10.

4. Clifford W. Brooks and Irvin Borish, *System for Ophthalmic Dispensing* (Oxford: Butterworth-Heinemann, 2006), p. 559.

5. Vinny R. Sastri, J. R. Perumareddi, and V. Ramachandra Rao, *Modern Aspects of Rare Earths and Their Complexes* (Amsterdam: Elsevier Science, 2003), p. 897.

CHAPTER 5. A SCIENTIFIC COLD WAR

1. Randall Bytrwerk, "Jerry Siegel Attacks!" German Propaganda Archive at Calvin College, http://www.calvin.edu/academic/cas/gpa/superman.htm (accessed April 11, 2014); "Jerry Siegel Attacks!" *Das Schwarze Korps*, April 25, 1940, p. 8.

2. Thomas Tuna, "In the Beginning . . . Gold," *Amazing World of DC Comics* 4, no. 16 (December 1971): 12–14; Wallace Harrington, "Superman and the War Years: The Battle of Europe within the Pages of Superman Comics," *Superman Homepage*, http://www.supermanhomepage.com/comics/comics.php?topic=articles/supes-war (accessed April 10, 2014).

3. John W. Poston, "Do Transuranic Elements Such as Plutonium Ever Occur Naturally?" *Scientific American*, March 23, 1998, http://www.scientificamerican.com/article/do-transuranic-elements-s/ (accessed January 24, 2014).

4. Sarah Charley, "How to Make an Element," *NOVA*, January 13, 2012, http://www.pbs.org/wgbh/nova/physics/make-an-element.html (accessed December 2, 2014).

5. "Reactor-Grade and Weapons-Grade Plutonium in Nuclear Explosives," *Nonproliferation and Arms Control Assessment of Weapons-Usable Fissile Material Storage and Excess Plutonium Disposition Alternatives* (Washington, DC: US Department of Energy Publications, 1997): 37–39.

6. Jonathan Tucker, *War of Nerves: Chemical Warfare from World War I to Al-Qaeda* (New York: Anchor, 2007), p. 16.

7. Ernest B. Hook, "Interdisciplinary Dissonance and Prematurity: Ida Noddack's Suggestion of Nuclear Fission," in *Prematurity in Scientific Discovery: On Resistance and Neglect*, ed. Ernest B. Hook (Berkeley: University of California Press, 2002), p. 133.

8. Emilio Segrè, *A Mind Always in Motion: The Autobiography of Emilio Segrè* (Berkeley: University of California Press, 1993), p. 115.

9. Robert E. Krebs, *The History and Use of Our Earth's Chemical Elements: A Reference Guide* (Westport, CT: Greenwood, 2006), p. 131.

10. Valerie Bailey Grasso, "Rare Earth Elements in National Defense: Background, Oversight Issues, and Options for Congress" (Congressional Research Service R41744, December 2013): 10–13.

CHAPTER 6. CREATED IN A NUCLEAR REACTOR

1. "Nuclear Power in France," World Nuclear Association, February 2014, http://www.world-nuclear.org/info/Country-Profiles/Countries-A-F/France/ (accessed March 29, 2014).

2. Albert Stwertka, *A Guide to the Elements*, 3rd ed. (Oxford: Oxford University Press, 2012), p. 192.

3. V. P. Visgin, "On the Origins of the Soviet Atomic Project: Role of the Intelligence Service 1941–1946," *Voprosy Istorii Estestvoznaniia i Tekhniki (Problems in the History of Science and Technology)*, no. 3 (1992): 97–134.

4. Gerald Ford, "Statement on Nuclear Policy," *American Presidency Project*, October 1976, http://www.presidency.ucsb.edu/ws/?pid=6561#axzz1zILTm1BT (accessed March 11, 2014).

5. Alina Selyukh, "Timeline of U.S. Nuclear Reprocessing," Reuters, August 17, 2010, http://www.reuters.com/article/2010/08/17/idINIndia-50883020100817 (accessed April 25, 2014).

6. "US Nuclear Recycling Faces the Axe," *Nature*, July 2, 2009 (accessed April 25, 2014); Eric Connor, "South Carolina Lawmakers Push Obama to Reconsider MOX Decision," *Greenville Online*, April 2, 2014, http://www.greenvilleonline.com/story/news/politics/make-your-voice-heard/2014/04/02/south-carolina-lawmakers-push-obama-to-reconsider-mox-decision/7190257/ (accessed April 25, 2014).

7. Mark Silva, "Obama's Manhattan Nuke Comment No Indication of Threat," Bloomberg, March 26, 2014, http://www.bloomberg.com/news/2014-03-26/obama-s-manhattan-nuke-comment-no-indication-of-threat.html (accessed April 24, 2014).

8. "The Use of Scientific and Technical Results from Underground Research Laboratory Investigations for the Geological Disposal of Radioactive Waste," *International Atomic Energy Agency* IAEA-TECDOC-1243 (September 2001): 2.

9. Gregory Benford et al., "Ten Thousand Years of Solitude? On Inadvertent Intrusion into the Waste Pilot Plant Repository," *Los Alamos National Laboratory* LA-12048-MS:1–31.

10. Joseph L. Kolb, "Inspectors Re-enter New Mexico Nuclear Waste Site after Leak," Reuters US, April 3, 2014, http://www.reuters.com/article/2014/04/03/usa-nuclear-newmexico-idUSL1N0MV00F20140403 (accessed April 23, 2014).

11. R. P. Bush, "Recovery of Platinum Group Metals from High Level Radioactive Waste," *Platinum Metals Review* 35, no. 4 (1991): 203.

12. P. Summanen et al., "Radiation Related Complications after Ruthenium Plaque Radiotherapy of Uveal Melanoma," *British Journal of Ophthalmology* 80, no. 8 (August 1996): 732–39.

13. Bush, "Recovery of Platinum Group Metals," pp. 202–208.

14. Ibid., p. 203.

15. Alex P. Meshik, "The Workings of an Ancient Nuclear Reactor," *Scientific American* (November 1, 2005): 82–91.

16. "Nature's Nuclear Reactors: The 2-Billion-Year-Old Natural Fission Reactors in Gabon, Western Africa," *Scientific American*, July 13, 2011, http://blogs.scientificamerican.com/guest-blog/2011/07/13/natures-nuclear-reactors-the-2-billion-year-old-natural-fission-reactors-in-gabon-western-africa/ (accessed October 18, 2013).

17. R. J. de Meijera, V. F. Anisichkin, and W. van Westrenen, "Forming the Moon from Terrestrial Silicate-Rich Material," *Chemical Geology* 345 (2013): 40–49.

CHAPTER 7. COUNTERFEITING GOLD

1. Gilbert King, "Fritz Haber's Experiments in Life and Death," *Smithsonian*, June 6, 2012, http://www.smithsonianmag.com/history/fritz-habers-experiments-in-life-and-death-114161301/?no-ist (accessed October 2, 2013).

2. Margaret MacMillan, "Ending the War to End All Wars," *New York Times*, December 25, 2010, http://www.nytimes.com/2010/12/26/opinion/26macmillan.html?pagewanted=all&_r=0 (accessed May 1, 2014).

3. Claire Suddath, "Why Did World War I Just End?" *Time*, October 4, 2010, http://content.time.com/time/world/article/0,8599,2023140,00.html (accessed May 2, 2014).

4. E. K. Holmyard, *Alchemy* (Mineola, NY: Dover, 1990), pp. 61–62, 69.

5. "Special Collections: Islamic Alchemy," Perfecting Nature: Medicine, Metallurgy, and Mysticism Exhibit, University of Delaware, http://www.lib.udel.edu/ud/spec/exhibits/alchemy/ia.html (accessed November 17, 2013).

6. K. Aleklett et al., "Energy Dependence of 209Bi Fragmentation in Relativistic Nuclear Collisions," *Physics Review C* 23 (1981): 1044–1046.

7. "Gold to Go Vending Machines," Gold Bars Worldwide, May 2013, http://www.goldbarsworldwide.com/PDF/RT_8_Gold_Vending_Machines.pdf (accessed May 2, 2014).

8. Felix Salmon, "The Problem of Fake Gold Bars," Reuters, March 25, 2012, http://blogs.reuters.com/felix-salmon/2012/03/25/the-problem-of-fake-gold-bars/ (accessed November 11, 2014); Rob Wile, "How a Manhattan Jeweler Wound Up with Gold Bars Filled with Tungsten," *Business Insider*, September 19, 2012, http://www.businessinsider.com/tungsten-filled-gold-bars-found-in-new-york-2012-9#ixzz30sF0UDk9 (accessed November 19, 2013).

9. Mike Duffy, "Counterfeit Aussie Gold Sold in China," Yahoo! News, October 22, 2012, https://au.news.yahoo.com/a/15182799/counterfeit-aussie-gold-sold-in-china/ (accessed May 5, 2014).

10. Ti-Hua Chang, "Fake Gold Bars Turn Up in Manhattan," MyFoxNY, October 9, 2012, http://www.myfoxny.com/story/19578206/fake-gold-bars-turn-up-in-manhattan (accessed November 19, 2013).

11. "Thermo Fisher Introduces Gold-Plate Detection Technology," Recycling Today, May 23, 2012, http://www.recyclingtoday.com/thermo-fisher-software-gold-detection.aspx (accessed May 5, 2014).

12. "Ultrasonic Testing of Gold Bars," Application Notes, Olympus, http://www.olympus-ims.com/en/applications/ut-testing-gold-bars/ (accessed May 6, 2014).

13. Evan Soltas, "Economics Is Platinum: What the Trillion-Dollar Coin Teaches Us," BloombergView, January 14, 2013, http://www.bloomberg.com/news/2013-01-14/economics-is-platinum-what-the-trillion-dollar-coin-teaches-us.html (accessed December 12, 2013).

14. "2012 Annual Report," United States Mint, 2012, http://www.usmint.gov/downloads/about/annual_report/2012AnnualReport.pdf (accessed December 12, 2013).

15. "Roman, British and Islamic Coin Collection Sells for £28,000," BBC News Sussex, January 24, 2014, http://www.bbc.com/news/uk-england-sussex-25881877 (accessed May 5, 2014)

16. James Grout, "Mark Antony and the Legionary Denarius," Roman Numismatics, Encyclopaedia Romana, http://penelope.uchicago.edu/~grout/encyclopaedia_romana/miscellanea/numismatics/antony.html (accessed May 5, 2014).

17. "Tungsten Alloy Scan Gold Coin," ChinaTungsten Online (Xiamen) Manufacturing and Sales Corporation, http://www.tungsten-alloy.com/tungsten-alloy-scan-gold-coin.html (accessed December 1, 2013).

CHAPTER 8. PALE HORSES

1. "BRITAIN: . . . Horseman, Pass By," *Time*, July 17, 1972, http://content.time.com/time/magazine/article/0,9171,877880,00.html (accessed November 18, 2013); Deborah Blum, "The Chemist as Murderer," *PLOS Blogs*, February 23, 2011, http://blogs.plos.org/speakeasyscience/2011/02/23/the-chemist-as-murderer/ (accessed September 12, 2013).

2. "Toxicology Review of Thallium and Compounds," Environmental Protection Agency EPA/635/R-08/001F, September 2009, p. 9.

3. A. Saha, "Erosion of Nails Following Thallium Poisoning: A Case Report," *Occupational and Environmental Medicine* 61 (2004): 640–42.

4. "Imprisonment for Mrs. Grills," *Canberra Times*, September 24, 1954, http://trove.nla.gov.au/ndp/del/article/2910501 (accessed April 17, 2014).

5. Blum, "The Chemist as Murderer."

6. Michael Howard, "Two Children Die as Iraqi Poison Plot Recalls Saddam's Assassination Method of Choice," *Guardian*, February 9, 2008, http://www.theguardian.com/world/2008/feb/09/iraq.international (accessed September 13, 2013).

7. Alan S. Cowell, *The Terminal Spy: A True Story of Espionage, Betrayal and Murder* (New York: Broadway, 2008) p. 265.

8. Christopher P. Holstege et al., "Thallium," in *Criminal Poisoning: Clinical and Forensic Perspectives* (Sudbury, MA: Jones & Bartlett Learning, 2010), p. 166.

9. "Radiation Emergencies & Response: Prussian Blue," Centers for Disease Control and Prevention, February 4, 2014, http://www.bt.cdc.gov/radiation/prussianblue.asp (accessed March 13, 2014).

10. "Radiogardase® (Prussian blue insoluble capsules)," Heyltex, http://www.heyltex.com/radiogardase.php (November 11, 2014).

11. "Who Was Alexander Litvinenko?" BBC News, July 12, 2013, http://www.bbc.com/news/uk-19647226 (article updated July 30, 2014 and title changed to "Alexander Litvinenko: Profile of Murdered Russian Spy" (accessed September 4, 2014).

12. "Russia Ready to Cooperate with Britain on Litvinenko Case," *Ria Novosti*, May 18, 2008, http://en.ria.ru/russia/20080518/107675252.html (accessed May 16, 2014).

13. Ryan Kisiel and Sam Greenhill, "Triple Agent! Poisoned Russian Spy Alexander Litvinenko Was Working for British AND Spanish Intelligence, Says Wife," *Daily Mail*, December14, 2012, http://www.dailymail.co.uk/news/article-2247486/Alexander-Litvinenko-Poisoned-Russian-spy-working-British-AND-Spanish-intelligence-says-wife.html (accessed May 16, 2014).

14. Richard Gray, "Litvinenko Waiter Recounts Polonium Poisoning," *Telegraph*, July 15, 2007, http://www.telegraph.co.uk/news/uknews/1557492/Litvinenko-waiter-recounts-polonium-poisoning.html (accessed October 21, 2013).

15. Robin B. McFee and Jerrod B. Leikin, "Death by Polonium-210: Lessons Learned from the Murder of Former Soviet Spy Alexander Litvinenko," *JEMS* 33, no. 20 (2008): 18–19.

16. Gray, "Litvinenko Waiter Recounts Polonium Poisoning."

17. Sandra Laville and Richard Norton-Taylor, "Litvinenko Was Victim of 'Russian Rogue Agents,'" *Guardian*, November 30, 2006, http://www.theguardian.com/uk/2006/dec/01/russia.politics (accessed October 25, 2013).

18. Anna Sadovnikova et al, "Litvinenko Mystery: 'Walking Dirty Bomb' Tells of London Meetings," *Spiegel Online International*, December 11, 2006, http://www.spiegel.de/international/spiegel/litvinenko-mystery-walking-dirty-bomb-tells-of-london-meetings-a-453803.html (accessed May 17, 2014).

19. Mark Oliver, "Lugovoi Extradition Refusal 'Disappointing,'" *Guardian*, July 10, 2007, http://www.theguardian.com/world/2007/jul/10/russia.politics (accessed May 17, 2014).

20. Patrice Mangin et al., "Expert Forensics Report Concerning the Late President Yasser Arafat," Centre hospitalier Universitaire Vaudois, November 5, 2013, http://static.guim.co.uk/ni/1383750812173/Yasser-Arafat-report.pdf (accessed January 3, 2014); "Factsheet on Forensics Report Concerning the Late President Yasser Arafat," Centre hospitalier Universitaire Vaudois, November 10, 2013, http://www.curml.ch/factsheet_arafat_201310081.pdf (accessed May 18, 2014), p. 1.

21. Saed Bannoura, "Abu Sharif: Israeli Mossad Poisoned Arafat through His Medications," Al Jazeera, July 20, 2009.

22. "Polonium-210," International Atomic Energy Agency, http://www.iaea.org/Publications/Factsheets/English/polonium210.html (accessed May 17, 2014).

23. Ibid.

24. "Toxic Fears Hit Highbury Auction," BBC Sport, May 10, 2006, http://news.bbc.co.uk/sport2/hi/football/teams/a/arsenal/4757797.stm (accessed October 11, 2013).

25. "'Record' Premier League Wage Bill," BBC News, May 28, 2008, http://news.bbc.co.uk/2/hi/7423254.stm (accessed October 11, 2013).

26. "McDonald's Pulls 'Shrek' Glasses for Cadmium," NBC News, June 4, 2010, http://www.nbcnews.com/id/37504287/ns/health-food_safety/t/mcdonalds-pulls-shrek-glasses-cadmium/#.U3S3JShRf9s (accessed December 17, 2013).

27. "What Diseases Are Associated with Chronic Exposure to Cadmium?" Agency for Toxic Substances & Disease Registry, May 12, 2008, http://www.atsdr.cdc.gov/csem/csem.asp?csem=6&po=12 (accessed May 16, 2014).

28. "Shrek Forever After," Box Office Mojo, http://www.boxofficemojo.com/movies/?id=shrek4.htm (accessed May 15, 2014).

29. L. O. Eboh and Boye E. Thomas, "Analysis of Heavy Metal Content in Cannabis Leaf and Seed Cultivated in Southern Part of Nigeria," *Pakistan Journal of Nutrition* 4, no. 5 (2005): 349–51.

30. Rebecca A. Alderden, "The Discovery and Development of Cisplatin," *Journal of Chemical Education* 83, no. 5 (2006): 728–34.

31. B. Rosenberg et al., "Inhibition of Cell Division in Escherichia Coli by Electrolysis Products from a Platinum Electrode," *Nature* 205 (1965): 689–99.

32. C. Kimme-Smith, "Mammograms Obtained with Rhodium vs Molybdenum Anodes: Contrast and Dose Differences," *American Journal of Roentgenology* 162, no. 6 (1994): 1313–317.

33. "FDA Backgrounder on Platinum in Silicone Breast Implants," US Food and Drug Administration, January 21, 2009, http://www.fda.gov/MedicalDevices/ProductsandMedicalProcedures/ImplantsandProsthetics/BreastImplants/UCM064040 (accessed January 23, 2014).

34. D. Krishnamurthy et al., "Comparison of High-Dose Rate Prostate Brachytherapy Dose Distributions with Iridium-192, Ytterbium-169, and Thulium-170 Sources," *Brachytherapy* 10, no. 6 (2011): 461–65.

35. G. Muto et al., "Thulium: Yttrium-Aluminum-Garnet Laser for En Bloc Resection of Bladder Cancer: Clinical and Histopathologic Advantages," *Urology* 83, no. 4 (2014): 851–55.

36. Roy Rushforth, "Palladium in Restorative Dentistry," *Platinum Metals Review* 48, no. 1 (2004): 30–31.

37. F. Eren, "The Effect of Erbium, Chromium: Yttrium-Scandium-Gallium-Garnet (Er,Cr:YSGG) Laser Therapy on Pain during Cavity Preparation in Paediatric Dental Patients: A Pilot Study.," *Oral Health Dental Management* 12, no. 2 (June 2013): 80–84.

CHAPTER 9. GOLF CLUBS, iPHONES, AND TRIBAL WARS

1. Joe Bavier, "Congo War-Driven Crisis Kills 45,000 a Month: Study," Reuters U.S., January 22, 2008, http://www.reuters.com/article/2008/01/22/us-congo-democratic-death-idUSL2280201220080122 (accessed May 24, 2014).

2. James Owen, "Mountain Gorillas Eaten by Congolese Rebels," *National*

Geographic News, January 19, 2007, http://news.nationalgeographic.com/news/2007/01/070119-gorillas.html (accessed October 5, 2013).

3. "Q&A: DR Congo's M23 Rebels," BBC News Africa, November 5, 2013, http://www.bbc.com/news/world-africa-20438531 (accessed May 25, 2014).

4. Pete Jones and David Smith, "Congo Rebels Take Goma with Little Resistance and to Little Cheer," *Guardian*, November 20, 2012, http://www.theguardian.com/world/2012/nov/20/congo-rebel-m23-take-goma (accessed on May 25, 2014).

5. "M23 Rebels Announce 'End of Rebellion' in DR Congo," France24, November 11, 2013, http://www.france24.com/en/20131105-drc-congo-m23-rebels-announce-end-of-rebellion-insurgency/ (accessed May 25, 2014).

6. "Poisoning the Poor—Electronic Waste in Ghana," Greenpeace International, August 5, 2008, http://www.greenpeace.org/international/en/news/features/poisoning-the-poor-electroni/ (accessed October 22, 2013).

7. Rebecca Keenan, "Australia Blocked Rare Earth Deal on Supply Concerns," Bloomberg, February 14, 2011, http://www.bloomberg.com/news/2011-02-14/australia-blocked-china-rare-earth-takeover-on-concern-of-threat-to-supply.html (accessed December 11, 2013).

8. Sarah-Jane Tasker, "Lynas Corporation Locks in Deal with Japan's Sojitz," *Australian*, March 31, 2011, http://www.theaustralian.com.au/business/lynas-corporation-locks-in-deal-with-japans-sojitz/story-e6frg8zx-1226030992776 (accessed May 20, 2014).

9. Keith Bradsher, "The Fear of a Toxic Rerun," *New York Times*, June 29, 2011, http://www.nytimes.com/2011/06/30/business/global/30rare.html?pagewanted=all (accessed December 11, 2013).

10. Stephanie Sta Maria, "Australian MP: Lynas Prohibited from Importing Waste," Yahoo! News Malaysia, September 16, 2012, https://my.news.yahoo.com/australian-mp-lynas-prohibited-importing-022124021.html (accessed December 11, 2013).

11. "SEC Adopts Rule for Disclosing Use of Conflict Minerals," US Securities and Exchange Commission, August 22, 2012, http://www.sec.gov/News/PressRelease/Detail/PressRelease/1365171484002#.U4DtmCiooTA (accessed October 3, 2013).

12. Brian X. Chen, "In E-Mail, Steve Jobs Comments on iPhone 4 Minerals," *Wired*, June 28, 2010 (accessed May 19, 2014).

13. Richard Padilla, "Apple Confirms Suppliers Do Not Use Unethically Sourced Tantalum," MacRumors, February 12, 2014, http://www.macrumors.com/2014/02/13/suppliers-do-not-use-conflict-mineral/ (accessed March 28, 2014).

CHAPTER 10. THE CONCENTRATION QUESTION

1. "100T Rotary Open Top Hopper," *BNSF Railway* 531300-531779, p. G51.

2. "Radioactive Elements in Coal and Fly Ash: Abundance, Forms, and Environmental Significance," *United States Geological Survey Fact Sheet* FS-163-97 (October 1997): 1–4.

3. Frank Reith et al., "Biomineralization of Gold: Biofilms on Gold," *Science* 313, no. 5784 (July 2006): 233–36; N. Wiesemann et al., "Influence of Copper Resistance Determinants on Gold Transformation by *Cupriavidus metallidurans* Strain CH34," *Journal of Bacteriology* 195, no. 10 (May 2013): 2298–2308.

4. Steven M. Stanley, *Earth System History*, 2nd ed. (New York: W. H. Freeman, 2004), p. 281.

5. J. R Butler and A. Z Smith, "Zirconium, Niobium and Certain Other Trace Elements in Some Alkali Igneous Rocks," *Geochimica et Cosmochimica Acta* 26, no. 9 (September 1962): 945–53.

6. "Rare Earth in Bayan Obo," Visible Earth catalog, NASA, July 2, 2001, http://visibleearth.nasa.gov/view.php?id=77723 (accessed May 25, 2014).

7. E. C. T. Chao et al., "Origin and Ages of Mineralization of Bayan Obo, the World's Largest Rare Earth Ore Deposit, Inner Mongolia, China," *United States Department of the Interior Open-File Report* 90-538 (1990): 1–6.

8. Yuan Zhongxix et al., "Geological Features and Genesis of the Bayan Obo REE Ore Deposit, Inner Mongolia, China," *Applied Geochemistry* 7, no. 5 (September 1992) 429–42.

9. Kui-Feng Yang et al., "Mesoproterozoic Carbonatitic Magmatism in the Bayan Obo Deposit, Inner Mongolia, North China: Constraints for the Mechanism of Super Accumulation of Rare Earth Elements," *Ore Geology Reviews* 40, no. 1 (September 2011): 122–31; Xiao-Yong Yang et al., "Geochemical Constraints on the Genesis of the Bayan Obo Fe–Nb–REE Deposit in Inner Mongolia, China," *Geochimica et Cosmochimica Acta* 73, no. 5 (March 2009): 1417–35.

10. M. P. Smith and P. Henderson, "Fractionation of the REE in a Carbonate Hosted Hydrothermal System: Bayan Obo, China," *Goldschmidt Conference Toulouse* (1998): 1421–22.

11. "Preferential Policies," *Baotou National Rare Earth Hi-Tech Industrial Development Zone: Investment Promotion Bureau*, http://www.rev.cn/en/pre.htm (accessed February 14, 2014).

12. Lindsey Hilsum, "Are Rare Earth Minerals Too Costly for Environment?" *PBS News Hour*, December 14, 2009, http://www.pbs.org/newshour/bb/asia/july-dec09/china_12-14.html (accessed January 28, 2014).

13. "Rare-Earth Mining in China Comes at a Heavy Cost for Local Vil-

lages," *Guardian Weekly*, August 7, 2012, http://www.theguardian.com/environment/2012/aug/07/china-rare-earth-village-pollution (accessed October 17, 2013).

14. Wayne M. Morrison and Rachel Tang, "China's Rare Earth Industry and Export Regime: Economic and Trade Implications for the United States" (Congressional Research Service R42510, April 2012): 8–13.

15. Keith Bradsher, "Chasing Rare Earths, Foreign Companies Expand in China," *New York Times*, August 24, 2011, http://www.nytimes.com/2011/08/25/business/global/chasing-rare-earths-foreign-companies-expand-in-china.html?pagewanted=all&_r=0 (accessed October 17, 2013).

16. Esther Tanquintic-Misa, "China Mulls Increasing Export Taxes of Precious Rare Earths," *International Business Times*, May 23, 2014, http://au.ibtimes.com/articles/553438/20140523/china-export-taxes-rare-earths-metals-minerals.htm#.U4KehSiooTA (accessed May 25, 2014).

17. Wieland Wagner, "The Limits of Reform: Deng Xiaoping's Legacy Divides Chinese Leadership," *Spiegel Online International*, October 12, 2007, http://www.spiegel.de/international/world/the-limits-of-reform-deng-xiaoping-s-legacy-divides-chinese-leadership-a-511160.html (accessed January 31, 2014).

18. Li Zhi-Sui, *The Private Life of Chairman Mao* (New York: Random House, 1996), p. 9.

19. Shaun Rein, "How to Fix Western-Chinese Relations," *Forbes*, December 14, 2010, http://www.forbes.com/2010/12/14/nobel-peace-prize-china-deng-gandhi-leadership-managing-rein.html (accessed February 11, 2014).

20. Robert Cottrell, "How Mrs. Thatcher Lost Hong Kong," *Independent*, August 30, 1992, http://www.independent.co.uk/arts-entertainment/how-mrs-thatcher-lost-hong-kong-ten-years-ago-fired-up-by-her-triumph-in-the-falklands-war-margaret-thatcher-flew-to-peking-for-a-lastditch-attempt-to-keep-hong-kong-under-british-rule--only-to-meet-her-match-in-deng-xiaoping-two-years-later-she-signed-the-agreement-handing-the-territory-to-china-1543375.html (accessed February 11, 2014).

21. Xinhua, "Film Makers Flock to Tractor Factory to Shoot Deng's Stories," *News Guandong*, July 26, 2004, http://www.newsgd.com/specials/deng100thbirthanniversary/newspictures/200407280046.htm (accessed February 11, 2014); Liang Quan, "When Ding Xiaoping Said," Chinese National Radio SRC-74, translated into English, August 16, 2007, http://nm.cnr.cn/nmzt/60dq/tjnmg/200704/t20070412_504442760.html (accessed February 11, 2014).

22. Evelyn Iritanz, "Great Idea but Don't Quote Him," *Los Angeles Times*, September 9, 2004, http://articles.latimes.com/2004/sep/09/business/fi-deng9 (accessed February 11, 2014).

23. Marla Cone, "Desert Lands Contaminated by Toxic Spills," *Los Angeles Times*, April 24, 1997, http://articles.latimes.com/1997-04-24/news/mn-51903_1_mojave-desert (accessed May 30, 2014).

24. Andrew Restuccia, "Troubled Mine Holds Hope for U.S. Rare Earth Industry," *Washington Independent*, October 25, 2010, http://washington independent.com/101462/california-mine-represents-hope-and-peril-for-u-s-rare-earth-industry (accessed May 30, 2014).

25. Richard Martin, "Molycorp's $1 Billion Rare-Earth Gamble," *CNN Money*, November 18, 2011, http://features.blogs.fortune.cnn.com/2011/11/18/molycorps-1-billion-rare-earth-gamble/ (accessed May 30, 2014).

CHAPTER 11. DIRTY RECYCLING

1. "How Much Gold Is in Goldschläger?" *Manhattan Gold & Silver*, December 7, 2010, http://www.mgsrefining.com/blog/post/2010/12/07/How-Much-Gold-Is-in-Goldschlager.aspx (accessed May 3, 2013).

2. C. James, "Thulium I," *Journal of the American Chemical Society* 33, no. 8 (August 1911): 1332–44.

3. "Waste Electrical and Electronic Equipment (WEEE) Directive," Conformance CE Marking & Product Safety, March 11, 2014, http://www.conformance.co.uk/adirectives/doku.php?id=weee (accessed May 25, 2014).

4. "Electronic Product Management Electronic Waste Recycling Fee," Cal-Recycle, January 11, 2013, http://www.calrecycle.ca.gov/electronics/act2003/retailer/fee/ (accessed September 20, 2013).

5. Mary Ann Cunningham and William P. Cunningham, "What a Long, Strange Trip It Has Been," *Principles of Environmental Science*, 2004, http://highered.mcgraw-hill.com/sites/0072919833/student_view0/chapter13/additional_case_studies.html (accessed May 26, 2014).

6. Xia Huo et al., "Elevated Blood Lead Levels of Children in Guiyu, an Electronic Waste Recycling Town in China," *Environmental Health Perspectives* 115, no. 7 (July 2007): 1113–17.

CHAPTER 12. AFGHANISTAN'S PATH TO PROSPERITY

1. Larry Goodson and Thomas H. Johnson, "Parallels with the Past—How the Soviets Lost in Afghanistan, How the Americans Are Losing," *Orbis* (Fall 2011): 586.

2. David N. Gibbs, "Afghanistan: The Soviet Invasion in Retrospect," *International Politics* 37 (June 2000): 237.

3. Ron Synovitz, "Afghanistan: Kabul Confirms New Effort to Buy Back U.S.-Built Stinger Missiles," *Radio Free Europe*, January 31, 2005, http://www.rferl.org/content/article/1057196.html (accessed February 18, 2014).

4. Alan Cullison and Yaroslav Trofimov, "Karzai Bans Ingredient of Taliban's Roadside Bombs," *Wall Street Journal*, February 3, 2010, http://online.wsj.com/news/articles/SB10001424052748703822404575019042216778962 (accessed February 20, 2014).

5. Maiwand Safi, "Explosives Fertilizer Ban Hits Afghan Fruit Growers," Environment News Service, June 6, 2011, http://ens-newswire.com/2011/06/07/explosives-fertilizer-ban-hits-afghan-fruit-growers/ (accessed February 11, 2014).

6. "World Development Indications," World Bank, http://data.worldbank.org/country/afghanistan (accessed February 20, 2014).

7. Steven B. Karch and Olaf Drummer, *Karch's Pathology of Drug Abuse*, 4th ed. (Boca Raton, FL: Chemical Rubber Company Press, 2008), p. 381.

8. Tom Vanden Brook, "Afghan Bomb Makers Shifting to New Explosives for IEDs," *USA Today*, June 25, 2013, http://www.usatoday.com/story/news/world/2013/06/25/ammonium-nitrate-potassium-chlorate-ieds-afghanistan/2442191/ (accessed February 20, 2014).

9. Ahmad Ramin Delasa, "Business Hurt by Anti-Pollution Drive in North Afghan City," Environment News Service, February 3, 2010, http://ens-newswire.com/2012/02/03/business-hurt-by-anti-pollution-drive-in-north-afghan-city/ (accessed February 20, 2014).

10. Andrew Scanlon, "Afghanistan, UNEP Launch USD $6 Million Initiative to Help Communities Adapt to Effects of Climate Change," United Nations Environment Program News Center, October 11, 2012, http://www.un.org/climatechange/blog/2012/10/11/afghanistan-unep-launch-usd-6-million-initiative-to-help-communities-adapt-to-effects-of-climate-change/ (accessed February 24, 2014).

11. Paul L. Schiff Jr., "Opium and Its Alkaloids," *American Journal of Pharmaceutical Education* 66 (Summer 2002): 186–94.

12. Graham Farrell and John Thorne, "Where Have All the Flowers Gone?: Evaluation of the Taliban Crackdown against Opium Poppy Cultivation in Afghanistan," *International Journal of Drug Policy* 16 (2005) 81–91; Pierre-Arnaud Chouvy, *Opium: Uncovering the Politics of the Poppy* (Cambridge, MA: Harvard University Press, 2010), p. 52.

13. Kathy Gannon, "After Record Opium Year, Afghans Plant New Poppy Crop," *Denver Post*, November 11, 2013, http://www.denverpost.com/nation-

world/ci_24519077/after-record-opium-year-afghans-plant-new-poppy (accessed February 9, 2014).

14. Raphael F. Perl, "Taliban and the Drug Trade," *CRS Reports for Congress* Order Code RS21041 (October 2001): CRS-3.

15. Jonathon Burch, "Afghanistan Now World's Top Cannabis Source: U.N.," Reuters US, March 31, 2010, http://www.reuters.com/article/2010/03/31/us-afghanistan-cannabis-idUSTRE62U0IC20100331 (accessed May 21, 2014).

16. James Risen, "U.S. Identifies Vast Mineral Riches in Afghanistan," *New York Times*, June 13, 2010, http://www.nytimes.com/2010/06/14/world/asia/14minerals.html?pagewanted=all&_r=0 (accessed February 18, 2014); Hamid Shalizi, "China's CNPC begins oil production in Afghanistan," Reuters, October 21, 2012, http://uk.reuters.com/article/2012/10/21/uk-afghanistan-oil-idUKBRE89K07Y20121021 (accessed February 18, 2014).

17. Zabihullah Ghazi, "Mountains of Wealth in Afghanistan," Environment News Service, May 23, 2013, http://ens-newswire.com/2013/05/23/mountains-of-wealth-in-afghanistan/ (accessed February 14, 2014).

18. Stephen G. Peters et al., "Preliminary Non-Fuel Mineral Resource Assessment of Afghanistan 2007," *United States Geological Survey Fact Sheet* 2007–3063 (October 2007): 1–4.

19. Stephen G. Peters, "Significant Potential for Undiscovered Resources in Afghanistan," United States Geological Survey Newsroom, November 13, 2007, http://www.usgs.gov/newsroom/article.asp?ID=1819#.U3zihyiooTA (accessed February 12, 2014).

CHAPTER 13. LITTLE SILVER

1. John B. Mertie Jr., "Economic Geology of the Platinum Metals," *Geological Survey Professional Paper* 630 (1969): 2.

2. Chris Barnard and Andrew Fones, "The Platinum Decathlon—A Tribute to the Foresight of Antoine Baumé," *Platinum Metals Review* 56, no. 3 (2012): 166.

3. Ian Freestone et al., "The Lycurgus Cup—A Roman Nanotechnology," *Cardiff Gold Bulletin* 40, no. 4 (2007): 272.

4. D. B. Harden and Jocelyn M. C. Toynbee, "The Rothschild Lycurgus Cup," *Archaeologia* 97 (1959): 179–212.

5. "Gold Nanoparticles: Properties and Applications," Sigma-Aldrich, http://www.sigmaaldrich.com/materials-science/nanomaterials/gold-nanoparticles.html (accessed May 3, 2014).

6. Sandra Davison and R. G. Newton, *Conservation and Restoration of Glass*, 2nd ed. (Oxford: Elsevier, 2003), p. 8.

7. L. B. Hunt, "The First Experiments on Platinum: Charles Wood's Samples from Spanish America," *Platinum Metals Review* 29, no. 4 (1985): 180–84.

8. Leslie B. Hunt and Donald McDonald, *A History of Platinum and Its Allied Metals* (London: Europa, 1982), pp. 147–50.

9. "History of Length Measurement," National Physical Laboratory (United Kingdom), http://www.npl.co.uk/educate-explore/posters/history-of-length-measurement/ (accessed May 30, 2014).

CHAPTER 14. THE NEXT PRECIOUS METALS

1. Tom Higham et al., "New Perspectives on the Varna Cemetery (Bulgaria)—AMS Dates and Social Implications," *Antiquity* 81 (2007): 640–54.

2. Douglass W. Bailey, *Balkan Prehistory: Exclusion, Incorporation and Identity* (London: Routledge, 2000), pp. 217–18.

3. Valerie Bailey Grasso, "Rare Earth Elements in National Defense: Background, Oversight Issues, and Options for Congress" (Congressional Research Service R41744, December 2013): 1.

4. "Report of Meeting Department of Defense Strategic Materials Protection Board," Department of Defense, The Strategic Materials Protection Board (December 2008): 5.

5. Ibid., p. 8.

6. Air Force Research Laboratory, "Defense Production Act Title III Project Establishes Domestic Source for Beryllium," Wright-Patterson Air Force Base, September 17, 2013, http://www.wpafb.af.mil/news/story.asp?id=123363533 (accessed April 11, 2014).

7. Steve Gorman, "As Hybrid Cars Gobble Rare Metals, Shortage Looms," Reuters US, August 31, 2009, http://www.reuters.com/article/2009/08/31/us-mining-toyota-idUSTRE57U02B20090831 (accessed December 5, 2013).

8. Stefanie Blendis, "Graphene: 'Miracle Material' Will Be in Your Home Sooner Than You Think," CNN, October 6, 2013, http://www.cnn.com/2013/10/02/tech/innovation/graphene-quest-for-first-ever-2d-material/ (accessed May 29, 2014).

9. Xiangfan Xu et al., "Length-Dependent Thermal Conductivity in Suspended Single-Layer Graphene," *Nature Communications* 5 (2014): 3689.

10. "Scientific Background on the Nobel Prize in Physics 2010," Royal Academy of Sciences, October 5, 2010, http://www.nobelprize.org/nobel

_prizes/physics/laureates/2010/advanced-physicsprize2010.pdf (accessed May 29, 2014).

11. Jae-Hyun Lee, "Wafer-Scale Growth of Single-Crystal Monolayer Graphene on Reusable Hydrogen-Terminated Germanium," *Science Express*, April 3, 2014, http://www.sciencemag.org/content/early/2014/04/02/science.1252268.abstract?sid=02ee6752-b76f-4f19-bfcb-af1eea94963c (accessed April 11, 2014).

12. David Cohen-Tanugi and Jeffrey C. Grossman, "Water Desalination across Nanoporous Graphene," *Nano Letters* 12, no. 7 (2012): 3602–3608; Anna Yu. Romanchuk et al., "Graphene Oxide for Effective Radionuclide Removal," *Physical Chemistry Chemical Physics* 15 (2013): 2321–27.

13. Richard Gray, "Graphene Used to Create More Pleasurable Condoms," *Telegraph*, November 20, 2013, http://www.telegraph.co.uk/science/science-news/10462976/Graphene-used-to-create-more-pleasurable-condoms.html (accessed May 29, 2014).

14. A. Paul Gill, "Lomiko's Graphene 3D Lab Files Patent for Multiple Material Printer Filament," *Lomiko Metals Press Release*, January 20, 2013, http://www.lomiko.com/public/files/news/LMR%20NR%20Graphene%203D%20files%20Patent%2001-20-2014.pdf (accessed May 29, 2014).

15. C. R. Hammond. "The Elements," *Handbook of Chemistry and Physics*, 81st ed. (Boca Raton, FL: CRC Press, 2004).

16. "Doramad Radioactive Zahncreme," Oak Ridge Allied Universities, February 17, 2009, http://www.orau.org/ptp/collection/quackcures/toothpaste.htm (accessed January 28, 2014).

17. "Report for the All Party Parliamentary Group on Thorium Energy Thorium-Fueled Molten Salt Reactors," Weinberg Foundation, June 2013, http://www.the-weinberg-foundation.org/wp-content/uploads/2013/06/Thorium-Fuelled-Molten-Salt-Reactors-Weinberg-Foundation.pdf (accessed January 5, 2014); "WASH-1097: The Use of Thorium in Nuclear Power Reactors," Brookhaven National Laboratory for the U.S. Atomic Energy Commission, (June 1969): 24.

18. "Ending the MSRE," Oak Ridge National Laboratory, http://web.ornl.gov/info/ridgelines/nov12/msre.htm (accessed April 29, 2014).

19. Murray W. Rosenthal, "An Account of Oak Ridge National Laboratory's Thirteen Nuclear Reactors," Oak Ridge National Laboratory and UT-Battelle, ORNL/TM-2009/181, March 2010, p. 24.

20. M. W. Rosenthal et al., "Molten Salt Reactors—History, Status, and Potential," *Nuclear Applications and Technology* 8 (1970).

21. Sylvain David et al., "Revisiting the Thorium-Uranium Nuclear Fuel Cycle," *Europhysics News* 38, no. 2 (March/April 2007): pp. 25–27.

22. Robert Hargraves, "Energy Cost Innovation, Part 1: Liquid Fuel Nuclear Reactors," *Energy Collective*, August 26, 2013, http://theenergycollective.com/roberthargraves/262916/energy-cost-innovation-liquid-fuel-nuclear-reactors (accessed January 3, 2014).

23. Ambrose Evans-Pritchard, "China Blazes Trail for 'Clean' Nuclear Power from Thorium," *Telegraph*, January 6, 2013, http://www.telegraph.co.uk/finance/comment/ambroseevans_pritchard/9784044/China-blazes-trail-for-clean-nuclear-power-from-thorium.html (accessed April 22, 2014).

24. "Thorium Reactors: Asgard's Fireball," *Economist*, April 12, 2014.

25. "The Nuke That Might Have Been," *Economist*, November 11, 2013, http://www.economist.com/blogs/babbage/2013/11/difference-engine-0 (accessed April 29, 2014).

26. "Niobium," Avalon Rare Metals, http://www.avalonraremetals.com/projects/target_commodities/rare_metals/niobium/ (accessed March 21, 2014).

27. Ian R. McNab, "Launch to Space with an Electromagnetic Railgun," *IEEE Transactions on Magnetics* 39, no. 1 (January 2003): 295–304.

28. Jeffrey Kluger, "South by Southwest: A Space Tourist Makes His Case," *Time*, March 12, 2013, http://science.time.com/2013/03/12/south-by-southwest-a-space-tourist-makes-his-case/ (accessed December 9, 2013).

29. Dave Drachlis, "Advanced Space Transportation Program: Paving the Highway to Space," NASA, www.nasa.gov/centers/marshall/news/background/facts/astp.html_prt.htm (accessed December 9, 2013).

30. Keith Veronese, "The Science of Rail Guns," *io9*, March 14, 2012, http://io9.com/5892516/the-science-of-rail-guns (accessed February 18, 2014).

31. W. L. Haisler et al., "Three-Dimensional Cell Culturing by Magnetic Levitation," *Nature Protocols* 8, no. 10 (2013): 1940–49.

32. Jade Boyd, "Scientists Grow Cells in 3-D Using Magnetic Fields," *LiveScience*, June 1, 2010, http://www.livescience.com/9929-scientists-grow-cells-3-magnetic-fields.html (accessed May 29, 2014).

33. C. B. Collins et al., "Accelerated Emission of Gamma Rays from the 31-yr Isomer of 178Hf Induced by X-Ray Irradiation," *Physical Review Letters* 82, no. 4 (1999): 695.

34. "JASON Defense Advisory Panel Reports," *Federation of American Scientists*, http://www.fas.org/irp/agency/dod/jason/ (accessed January 22, 2014).

35. Ann Finkbeiner, "DARPA and Jason Divorce," *Science*, March 25, 2002, http://news.sciencemag.org/2002/03/darpa-and-jason-divorce (accessed January 23, 2014).

36. N. Lewis et al., "High Energy Density Explosives," JASON and Mitre Corporation, JSR-97-110, October 8, 1997, p. 13.

37. Andy Oppenheimer, "Mini-Nukes: Boom or Bust?" *Bulletin of the Atomic Scientists* 60, no. 5 (September/October 2004): 13.

38. Peter D. Zimmerman, "The Strange Tale of the Hafnium Bomb: A Personal Narrative," *APS Physics*, www.aps.org/publications/apsnews/200706/back page.cfm (accessed January 24, 2014).

39. Duncan Graham-Rowe, "Nuclear-Powered Drone Aircraft on Drawing Board," *New Scientist*, February 19, 2003, http://www.newscientist.com/article/dn3406#.U3ETFCi9ZdE (accessed May 12, 2014).

CHAPTER 15. WHEN THE WELL RUNS DRY

1. Jared M. Diamond, *Guns, Germs, and Steel: The Fates of Human Societies* (New York City: W.W. Norton & Co., 1999), p. 44.

2. "The Protocol on Environmental Protection to the Antarctic Treaty," Secretariat of the Antarctic Treaty, 2011, http://www.ats.aq/e/ep.htm (accessed May 27, 2014).

3. "Convention on the Regulation of Antarctic Mineral Resource Activities," *New Zealand Ministry of Foreign Affairs and Trade*, http://www.mfat.govt.nz/Treaties-and-International-Law/01-Treaties-for-which-NZ-is-Depositary/0-Antarctic-Mineral-Resource.php (accessed September 17, 2014).

4. "The World Factbook: Antarctica," Central Intelligence Agency, March 27, 2014, https://www.cia.gov/library/publications/the-world-factbook/geos/ay.html (accessed May 27, 2014).

5. "Treaty Partners," Australian Government Department of the Environment, April 4, 2012, http://www.antarctica.gov.au/law-and-treaty/treaty-partners (accessed May 27, 2014).

6. "Antarctic Territorial Claims," Australian Government Department of the Environment, August 12, 2010, http://www.antarctica.gov.au/law-and-treaty/treaty-partners (accessed May 27, 2014).

7. "Opportunities Overview," United States Antarctic Program, http://www.usap.gov/jobsAndOpportunities/ (accessed May 27, 2014).

8. Alister Doyle, "Antarctica May Have a New Type of Ice: Diamonds," Reuters US, December 17, 2013, http://www.reuters.com/article/2013/12/17/us-antarctica-diamonds-idUSBRE9BG0XF20131217 (accessed May 27, 2014).

9. D. Dick et al., "Rare Earth Elements Determined in Antarctic Ice by Inductively Coupled Plasma—Time of Flight, Quadrupole and Sector Field-Mass Spectrometry: An Inter-Comparison Study," *Analytica Chimica Acta* 621, no. 2 (July 2008): 140–47.

10. "Greenland Votes to Allow Uranium, Rare Earths Mining," Reuters, October 25, 2013, http://uk.reuters.com/article/2013/10/25/greenland-uranium-idUKL5N0IE4MJ20131025 (accessed December 7, 2013).

11. Peter Levring, "Greenland End to 25-Year Uranium Mining Ban Gets Danish Backing," Bloomberg News, October 25, 2013, http://www.bloomberg.com/news/2013-10-25/greenland-end-to-25-year-uranium-mining-ban-gets-danish-backing.html (accessed December 7, 2013).

12. David Michelsen, ed., "Greenland in Figures, 2013," *Statistics Greenland*, no. 10 (March 2013): 4–6.

13. Ibid., p. 7.

14. Adam Currie, "Rare Earth Mining in Greenland," *Resource Investing News*, September 3, 2012, http://resourceinvestingnews.com/42337-rare-earth-mining-in-greenland.html (accessed November 10, 2013).

15. David McFadden, "Official: Rare-Earth Elements in Jamaica's Red Mud," Associated Press, *The Big Story*, January 15, 2013, http://bigstory.ap.org/article/official-rare-earth-elements-jamaicas-red-mud (accessed January 11, 2014).

16. Eliane Cristina De Resende, "Synergistic Co-Processing of Red Mud Waste from the Bayer Process and a Crude Untreated Waste Stream from Bio-Diesel Production," *Green Chemistry* 15, no. 2 (2013): 496–510.

17. Willard Pinnock and Arun Wagh, "Occurrence of Scandium and Rare Earth Elements in Jamaican Bauxite Waste," *Economic Geology* 82 (1987): 757–61.

18. Colleen Hanahan, "Chemistry of Seawater Neutralization of Bauxite Refinery Residues (Red Mud)," *Environmental Engineering Science* 21, no. 2 (March 2004): 125–38; S. Rai et al., "Feasibility of Red Mud Neutralization with Seawater Using Taguchi's Methodology," *International Journal of Environmental Science and Technology* 10, no. 2 (March 2013): 305–14.

19. David Gura, "Toxic Red Sludge Spill from Hungarian Aluminum Plant 'An Ecological Disaster,'" NPR, *The Two-Way*, October 5, 2010, http://www.npr.org/blogs/thetwo-way/2010/10/05/130351938/red-sludge-from-hungarian-aluminum-plant-spill-an-ecological-disaster (accessed May 27, 2014); "Hungarian Chemical Sludge Spill Reaches Danube," BBC News Europe, October 7, 2010, http://www.bbc.co.uk/news/world-europe-11491412 (accessed October 11, 2013); Alan Taylor, "A Flood of Red Sludge, One Year Later," *Atlantic*, September 28, 2011, http://www.theatlantic.com/infocus/2011/09/a-flood-of-red-sludge-one-year-later/100158/ (accessed May 27, 2014).

20. "PM Simpson Miller Breaks Ground for 'Historic' Red Mud Pilot Plant Project," Office of the Prime Minister of Jamaica, February 5, 2013, http://opm.gov.jm/pm-simpson-miller-breaks-ground-for-historic-red-mud-pilot-plant-project/ (accessed May 27, 2014).

21. John D. Durand, "The Population Statistics of China, AD 2-1953," *Population Studies* 13, no. 3 (1960): 209–56.

22. Matt Schiavenza, "How Humiliation Drove Modern Chinese History," *Atlantic*, October 25, 2013, http://www.theatlantic.com/china/archive/2013/10/how-humiliation-drove-modern-chinese-history/280878/ (accessed May 28, 2014).

23. Keith Veronese, "During World War II, Japan Plotted to Unleash a Plague on the United States," *io9*, May 10, 2012, http://io9.com/5908290/during-world-war-ii-japan-plotted-to-unleash-a-plague-on-the-united-states (accessed October 11, 2013).

24. "China Claims It Agreed with Japan to Shelve the Dispute in 1972, Japan Denies," *The Asahi Shimbun*, December 26, 2012, https://ajw.asahi.com/article/special/Senkaku_History/AJ201212260103 (accessed February 18, 2014).

25. Aibing Guo and Rakteem Katakey, "Disputed Islands With 45 Years of Oil Split China, Japan," Bloomberg News, October 11, 2012, http://www.bloomberg.com/news/2012-10-10/disputed-islands-with-45-years-of-oil-split-china-japan.html (accessed February 18, 2014).

26. Benjamin Kang Lim, "Exclusive: China Tried to Convince North Korea to Give Up Nuclear Tests," Reuters US, June 4, 2013, http://www.reuters.com/article/2013/06/04/us-korea-north-china-idUSBRE95305H20130604 (accessed February 19, 2014).

27. "Polymetallic Nodules," International Seabed Authority, 2013, http://www.isa.org.jm/files/documents/EN/Brochures/ENG7.pdf (accessed November 29, 2013).

28. X. Wang, "Molecular Biomineralization: Toward an Understanding of the Biogenic Origin of Polymetallic Nodules, Seamount Crusts, and Hydrothermal Vents," *Progress in Molecular and Subcellular Biology* 52 (2011): 77–110; Peter Balaz and Jozef Franzen, "Rare Earth Elements in the Polymetallic Nodules—A New Challenge," *Proceedings of the Twenty-Second International Offshore and Polar Engineering Conference* (June 2012): 112–16.

29. "Polymetallic Nodules."

30. Ibid.

31. Trent Schindler, "Mysteries of the Deep: Raising Sunken Ships," Public Broadcasting Service, http://www.pbs.org/saf/1305/features/ship.htm (accessed May 28, 2014).

32. James Phelan, "An Easy Burglary Led to the Disclosure of Hughes-C.I.A. Plan to Salvage Soviet Sub," *New York Times*, March 27, 1975, http://select.nytimes.com/gst/abstract.html?res=FB0716FF395E157493C5AB1788D85F418785F9&scp=3&sq=Glomar%20Hughes&st=cse (accessed December 2, 2013).

33. "Russia Unveils the Mystery of Sunken K-129 Submarine," *Pravda*, September 10, 2007, http://english.pravda.ru/news/russia/10-09-2007/96959-sunken_submarine-0/#.U4YEHiiooTA (accessed May 28, 2014).

34. "Glomar Explorer Spies on Seabed Minerals," *New Scientist* 79, no. 1120 (September 14, 1978): 748–49.

35. "Marine Mining," De Beers, 2014, http://www.debeersgroup.com/en/explore-de-beers/mining.html (accessed September 22, 2014).

36. Gerald Traufetter, "Underwater Resources: Treasure at the Bottom of the Sea," *Spiegel Online International*, October 4, 2006, http://www.spiegel.de/international/spiegel/underwater-resources-treasure-at-the-bottom-of-the-sea-a-440618.html (accessed May 29, 2014).

37. "Marine Mining."

38. "Genesis and Occurrence of Deep Sea Mineral Deposits (Polymetallic Nodules)," CSIR, National Institute of Oceanography, http://www.nio.org/?option=com_projectdisplay&task=view&tid=2&sid=15&pid=15 (accessed May 29, 2014); T. V. Padma and Paula Park, "India Joins Deep Sea Mining Race," *Guardian*, August 30, 2012, http://www.theguardian.com/environment/2012/aug/30/india-mining (accessed May 29, 2014).

39. Keith Bradsher, "Amid Tension, China Blocks Vital Exports to Japan," *New York Times*, September 22, 2010, http://www.nytimes.com/2010/09/23/business/global/23rare.html?pagewanted=all&_r=0 (accessed May 28, 2014); Tania Branigan and Justin McCurry, "Japan Releases Chinese Fishing Boat Captain," *Guardian*, September 24, 2010, http://www.theguardian.com/world/2010/sep/24/japan-free-chinese-boat-captain (accessed May 28, 2014).

40. Kristina Hughes and Tom Miles, "China Loses Trade Dispute over Rare Earth Exports," Reuters US, http://www.reuters.com/article/2014/03/26/us-china-wto-rareearths-idUSBREA2P0ZK20140326 (accessed May 28, 2014).

CHAPTER 16. GOING THE DISTANCE

1. James Fincannon, "Six Flags on the Moon: What Is Their Current Condition?" *Apollo Lunar Surface Journal*, April 21, 2012, http://www.hq.nasa.gov/alsj/ApolloFlags-Condition.html (accessed November 11, 2013).

2. "Agreement Governing the Activities of States on the Moon and Other Celestial Bodies," United Nations Office for Disarmament Affairs, http://disarmament.un.org/treaties/t/moon (accessed March 20, 2014).

3. Sherree Owens Zalampas, *Adolf Hitler: A Psychological Interpretation of His Views on Architecture, Art, and Music* (Bowling Green, OH: Bowling Green Uni-

versity Popular Press, 1990), p. 57; Virgiliu Pop, "The Men Who Sold the Moon: Science Fiction or Legal Nonsense?" *Space Policy* 17, no. 3 (2001): 195–203.

4. "3 Yemenis Sue NASA for Trespassing on Mars," CNN, July 24, 1997, http://www.cnn.com/TECH/9707/24/yemen.mars/ (accessed September 17, 2014); "Yemenis Claim Mars," BBC News, March 22, 1998, http://news.bbc.co.uk/2/hi/despatches/67814.stm (accessed December 12, 2013).

5. Rachel Hardwick, "Dennis M. Hope Has Owned the Moon Since 1980 Because He Says So," *Vice*, April 11, 2013, http://www.vice.com/read/ive-owned-the-moon-since-1980 (accessed December 12, 2013).

6. Kenneth Chang, "After 17 Years, a Glimpse of a Lunar Purchase," *New York Times*, March 31, 2010, http://www.nytimes.com/2010/03/31/science/space/31moon.html?ref=science&_r=0 (accessed December 13, 2013).

7. "Luna 21/Lunokhod 2," National Space Science Data Center, NASA, http://nssdc.gsfc.nasa.gov/nmc/spacecraftDisplay.do?id=1973-001A (accessed December 13, 2013); "Western Researcher Solves 37-Year-Old Space Mystery," *Western News*, March 16, 2010, http://communications.uwo.ca/com/western_news/stories/western_researcher_solves_37-year-old_space_mystery_20100316445952/ (accessed December 14, 2013).

8. "U.S.S.R. Luna 21 Lander," NASA, March 22, 2010, http://www.nasa.gov/mission_pages/LRO/multimedia/lroimages/lroc-20100322-luna21.html (accessed May 30, 2014).

9. Tim Morrison, "Q & A with Space Tourist Richard Garriott," *Time*, September 24, 2008, http://content.time.com/time/health/article/0,8599,1844160,00.html (accessed May 1, 2014).

10. Tariq Malik, "Vintage Soviet Space Capsule Sold for Record $2.9 Million," Space.com, April 13, 2011, http://www.space.com/11377-soviet-space-capsule-sothebys-auction.html (accessed December 14, 2013).

11. Stephen P. Maran and Laurence A. Marschall, *Galileo's New Universe: The Revolution in Our Understanding of the Cosmos* (Dallas: BenBella, 2009), p. 41.

12. T. D. Lin et al., "Physical Properties of Concrete Made with Apollo 16 Lunar Soil Sample," 2nd Conference on Lunar Bases and Space Activities, 1992, 483–87.

13. Matthias Willbold et al., "The Tungsten Isotopic Composition of the Earth's Mantle before the Terminal Bombardment," *Nature* 477 (2011): 195–98.

14. Jessica Marshall. "Meteorites Pummeled Earth, Delivering Gold," *Discovery News*, September 7, 2011, http://news.discovery.com/space/astronomy/earth-meteorites-gold-metals-110907.htm (accessed February 4, 2014).

15. Nancy Atkinson, "The Most Profitable Asteroid Is . . ." *Universe Today*, May 16, 2012, http://www.universetoday.com/95169/the-most-profitable-asteroid-is/ (accessed February 4, 2014).

16. "About Asteroid Explorer "HAYABUSA" (MUSES-C)," Japan Aerospace Exploration Agency, June 26, 2013, http://global.jaxa.jp/projects/sat/muses_c/ (accessed January 27, 2014).

17. "Hayabusa Landed on and Took off from Itokawa Successfully—Detailed Analysis Revealed," Institute of Space and Aeronautical Science, Japan Aerospace Exploration Agency, November 24, 2005, http://www.isas.jaxa.jp/e/snews/2005/1124_hayabusa.shtml (accessed December 2, 2013).

18. Takanao Saiki et al., "Small Carry-On Impactor of Hayabusa2 Mission," *Acta Astronautica* 84 (2013): 227.

19. Mary L. Guerrieri, ed., *Resources of Near-Earth Space* (Tucson: University of Arizona Press, 1994) pp. 493–522.

20. "New NASA Mission to Help Us Learn How to Mine Asteroids," NASA, August 8, 2013, http://www.nasa.gov/content/goddard/new-nasa-mission-to-help-us-learn-how-to-mine-asteroids/ (accessed February 25, 2014).

2. William A. Ambrose et al., ed., *Energy Resources for Human Settlement in the Solar System and Earth's Future in Space* (Tulsa, OK: American Association of Petroleum Geologists, 2013), p. 84.

22. J. R. Olds et al., "Multiple Mass Drivers as an Option for Asteroid Deflection Missions," *Proceedings of Planetary Defense Conference* (2007): 4–23.

23. "NASA—Advanced Space Transportation Program Fact Sheet," Advanced Space Transportation Program, NASA, http://www.nasa.gov/centers/marshall/news/background/facts/astp.html_prt.htm (accessed April 2, 2014).

INDEX